AF882769

Preschool Math Concepts

For Beginner Learners

ISBN: 978-1-7353983-1-0

www.printingpublishingpalace.com

name

is
Super Smart!

Zero

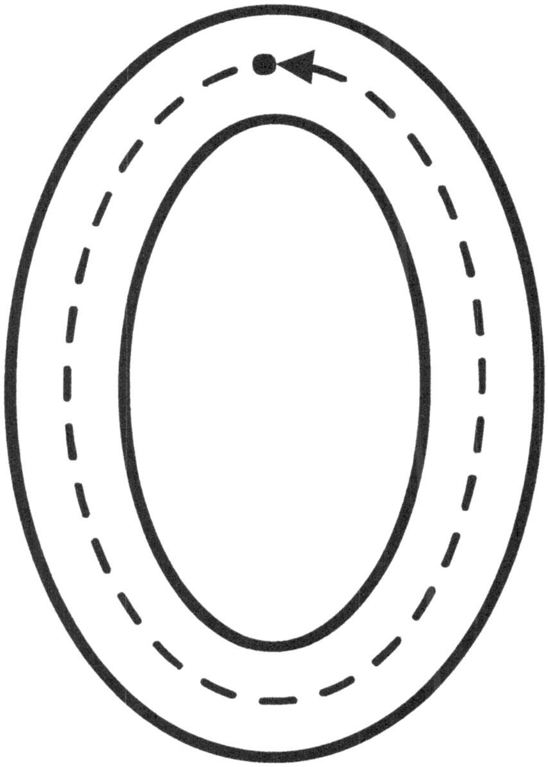

Circle the number 0

3 0 1

2 3

7

0 5

6 4

8 0

1 7

10 5

2

0 9

7 0

Color the shape blue with Zero sides

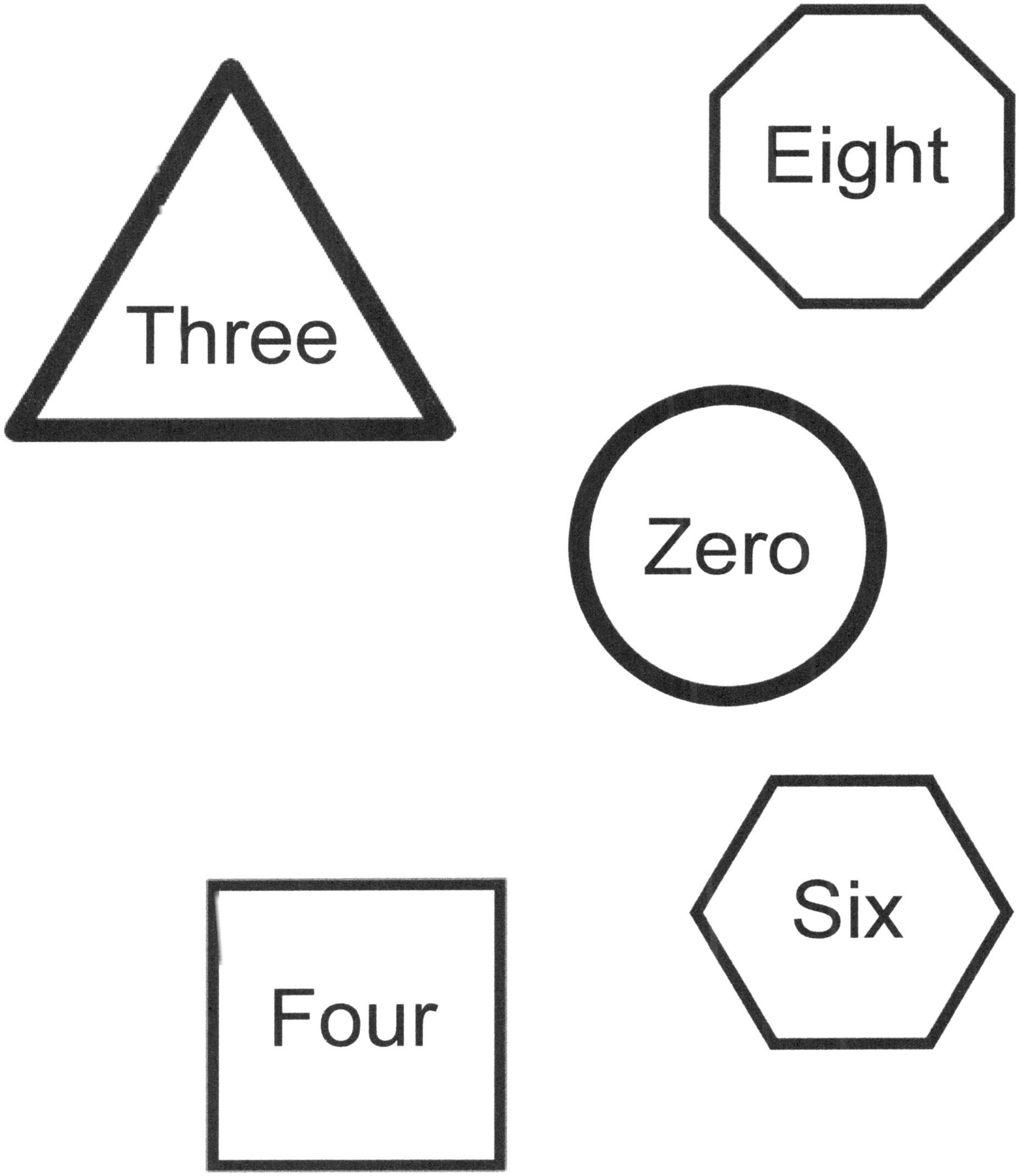

Domino Addition

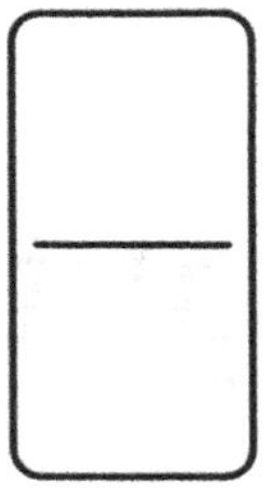 + 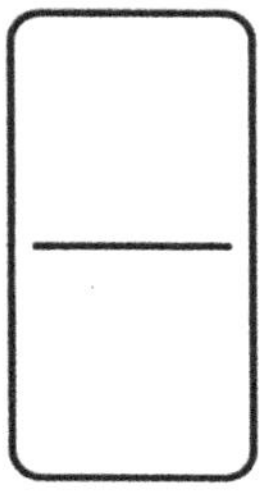 = ______

0 ______ 0 ______

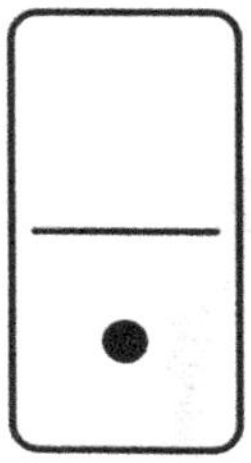 + 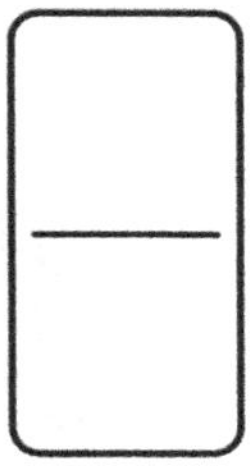 = ______

______ ______

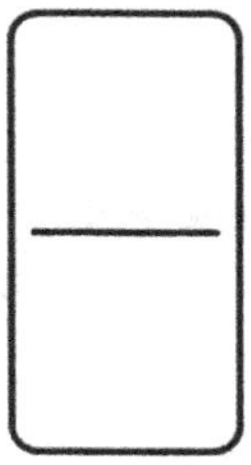 + 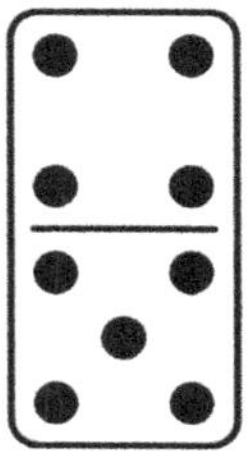 = ______

______ ______

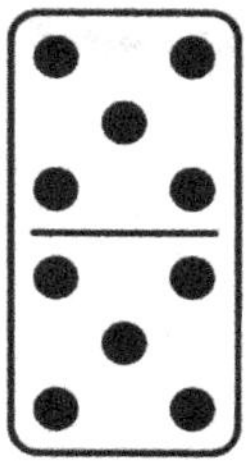 + 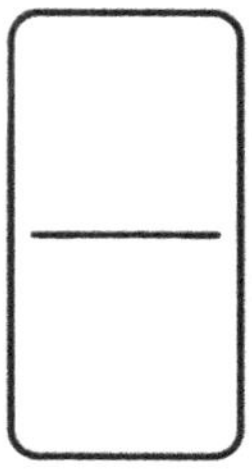 = ______

______ ______

Subtraction

 -  = ______

1 ______ 1 ______

 -  = ______

______ ______

 -  = ______

______ ______

 - 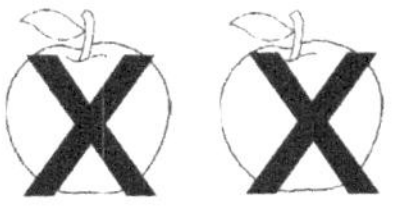 = ______

______ ______

Draw a line from the basket to every 0

10

0

7

5

0

6 6 9 4

2 0 8 6

5 1 0 5

3

0

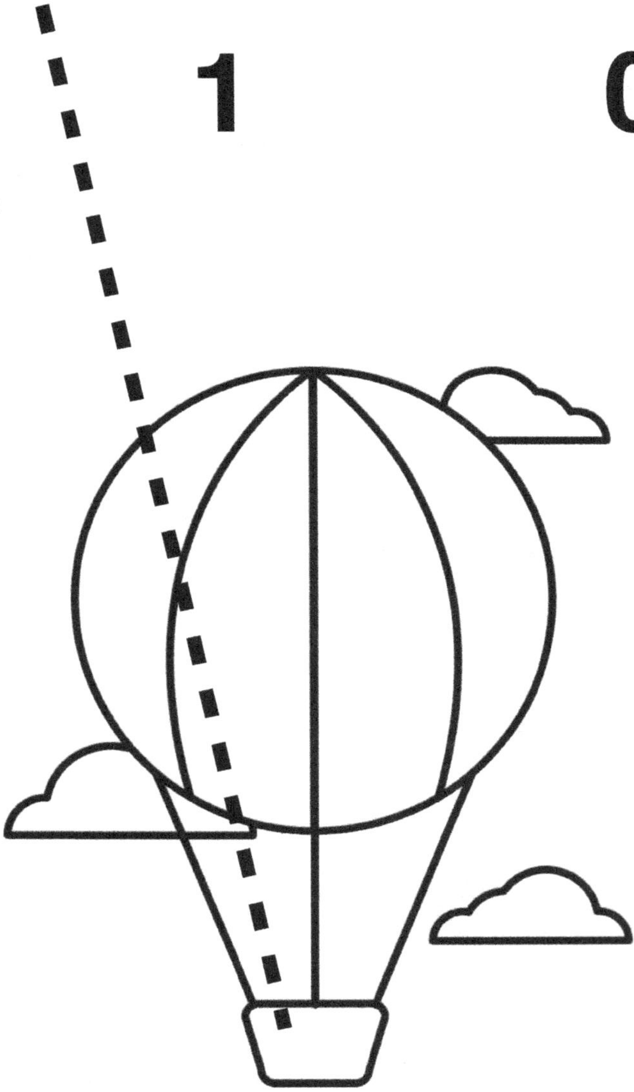

One

Use red to color the number 1

3 2 4 1

3 1 5 7

6 1

8 1

1 7

2 1 5

1 1 9

7

Draw a line to show how many balloons are there

2

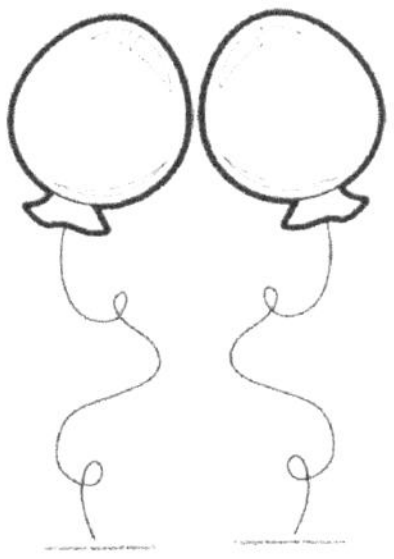

1

2

1

6

2

Domino Addition

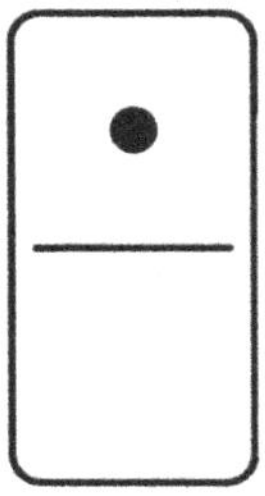 + 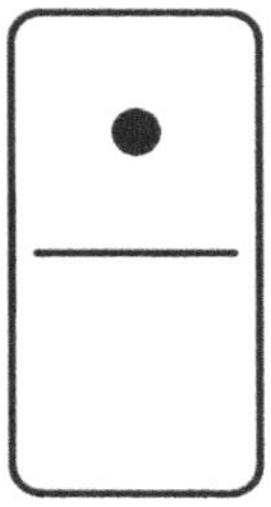 = ____

1   1

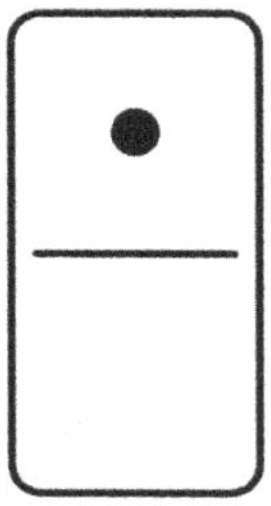 + 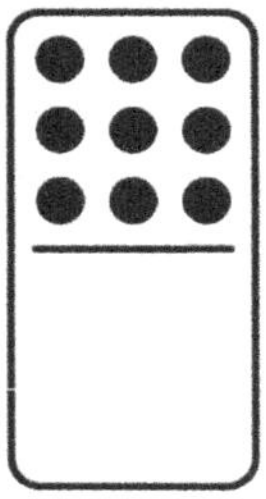 = ____

____   ____

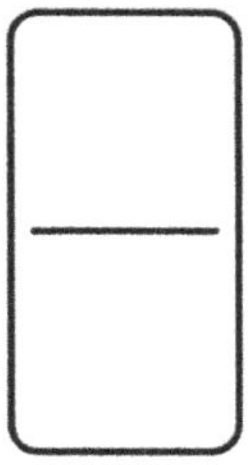 + 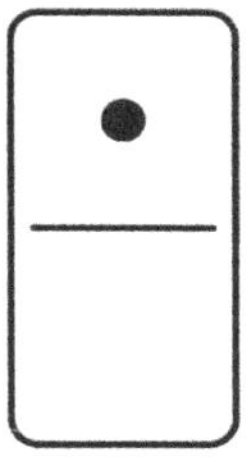 = ____

____   ____

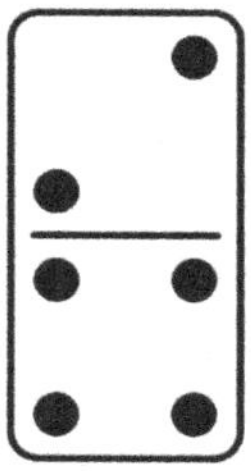 + 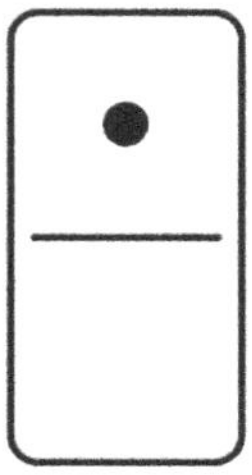 = ____

____   ____

Subtraction

 -  = ______

4 ______ 1 ______

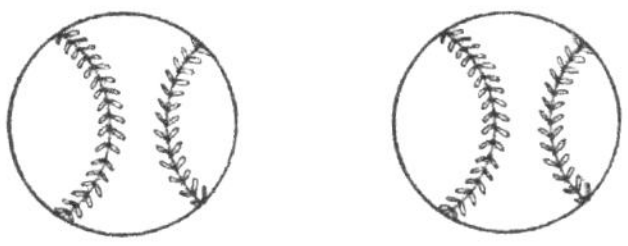 -  = ______

______ ______

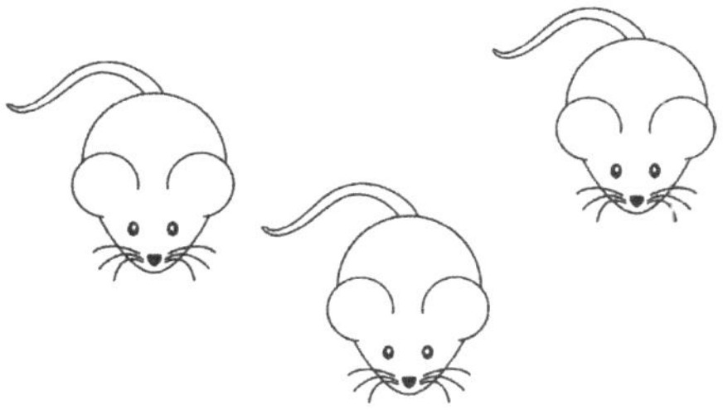 -  = ______

______ ______

 - 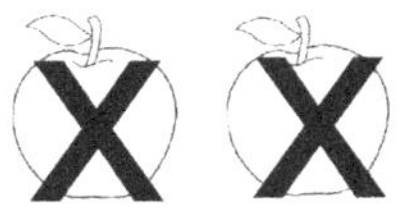 = ______

______ ______

Top or Bottom

Underline the number on the **top**

One

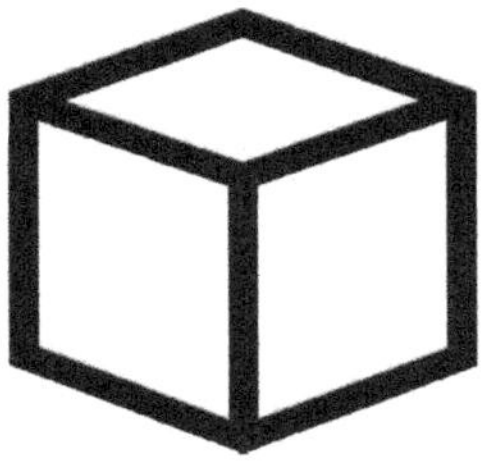

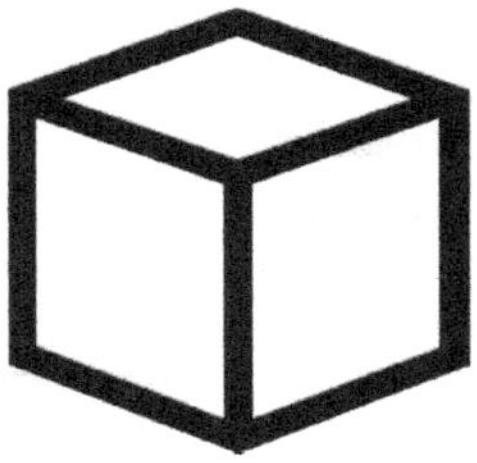

Five

Underline the number on **the bottom**

0

1

Circle the numbers on the **top**

0, 1, 2, 3, 4, 5

6, 7, 8, 9, 10

Circle the numbers on the **bottom**

10, 9, 8, 7, 6

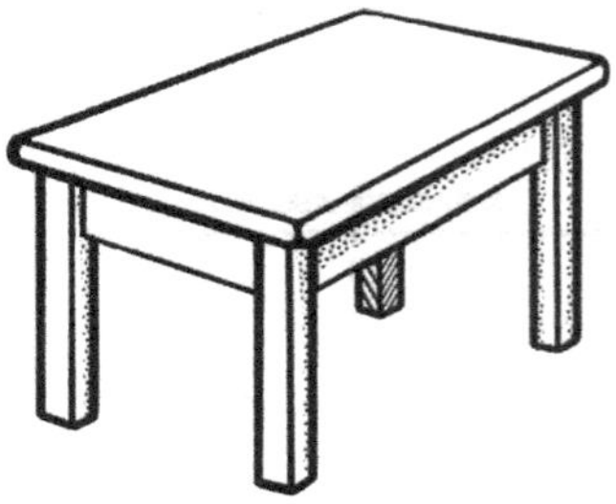

5, 4, 3, 2, 1, 0

Color the number **on the top**

Color the number on the **bottom**

Two

Color each square with the number 2

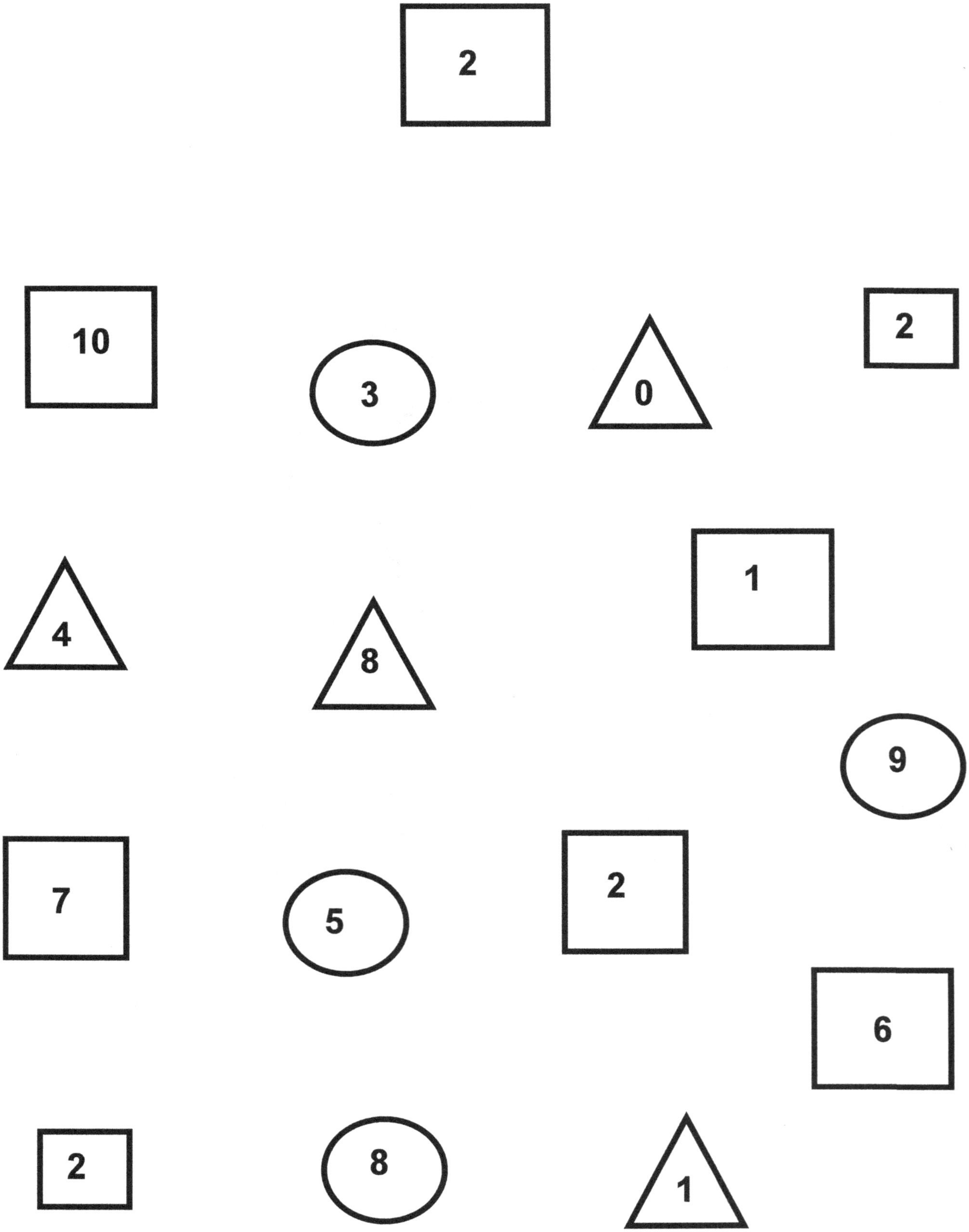

Color the houses with **a two number on the door**

Domino Addition

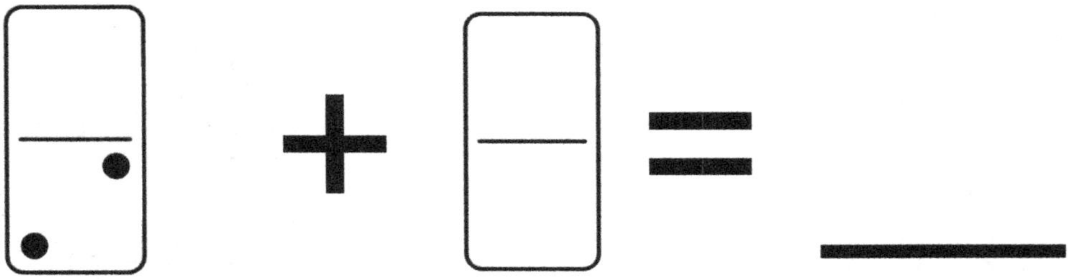

2 0

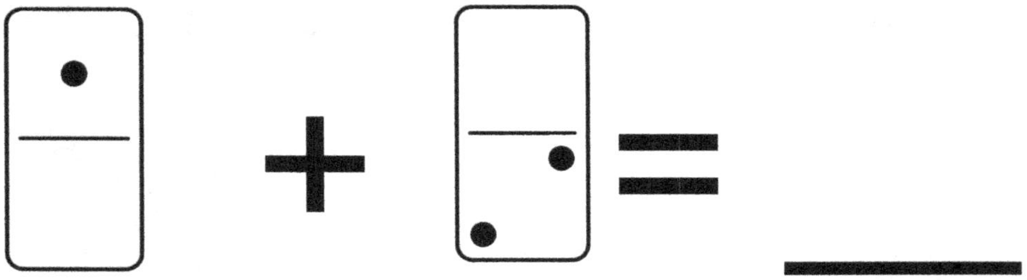

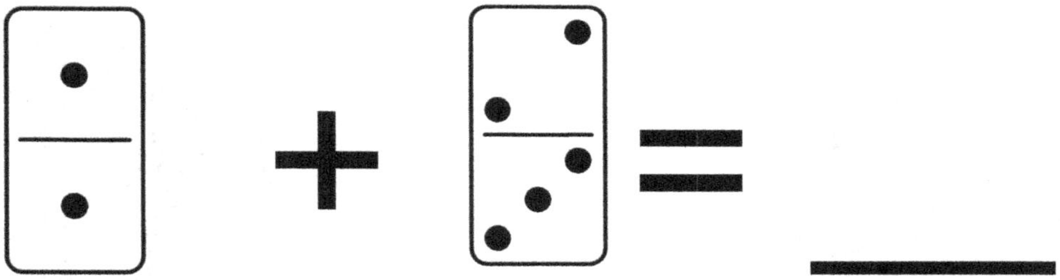

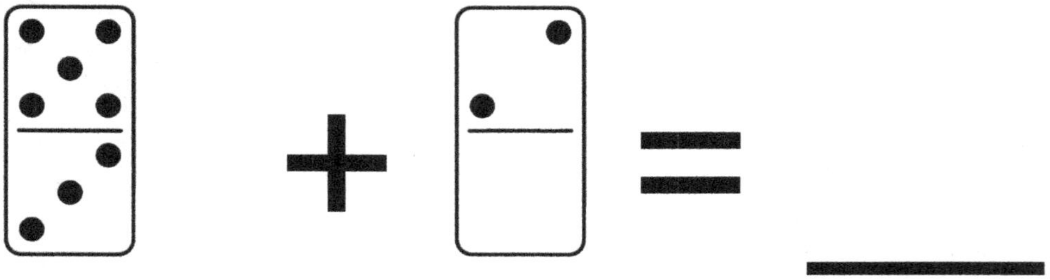

Subtraction

2 1

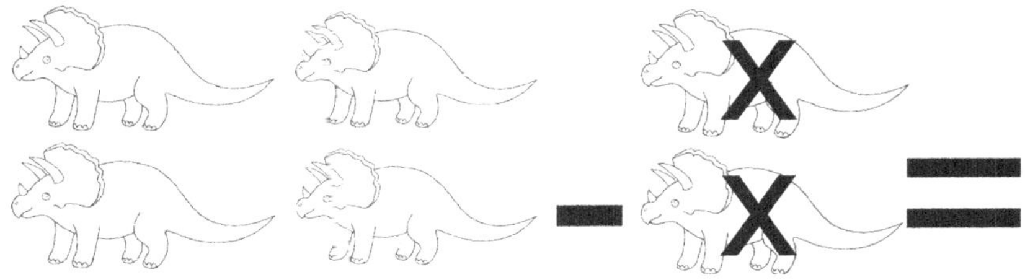

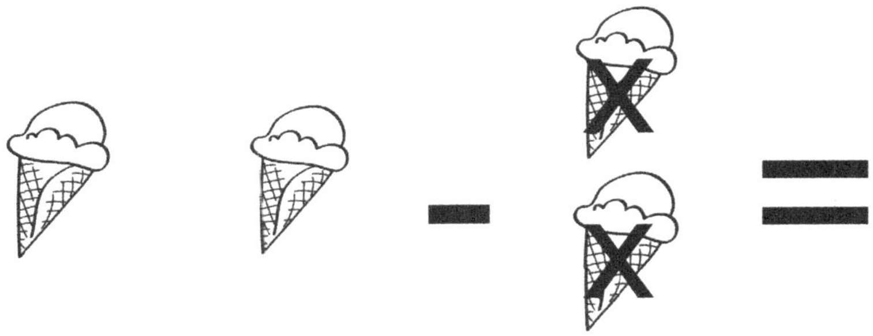

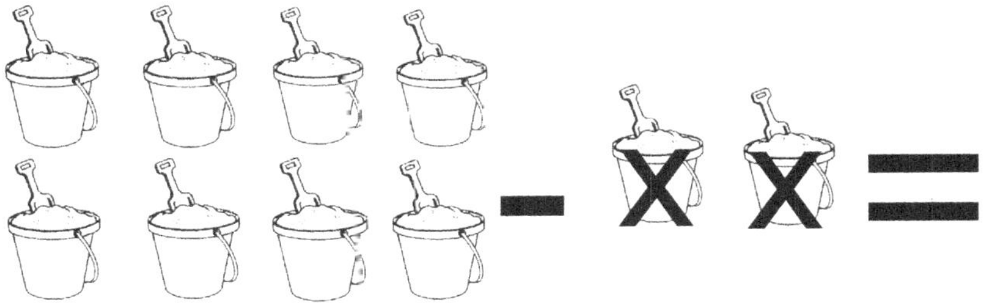

Draw a line from the photos with 2 birds to the number two

Three

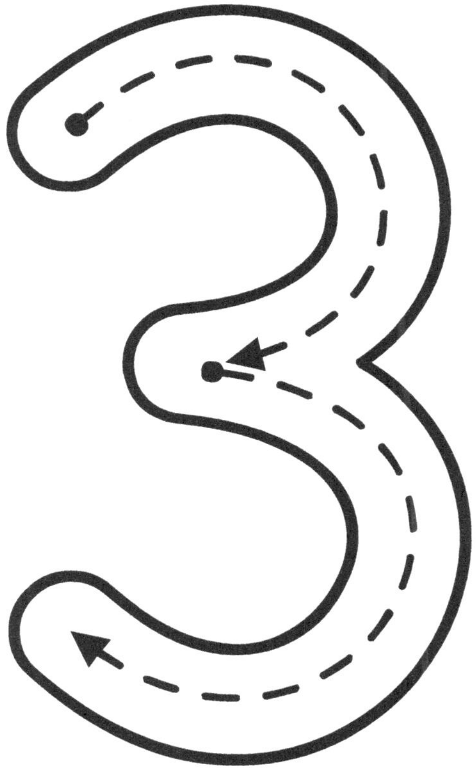

Count and underline the correct number

	10 2 3
	5 7 1
	1 0 5
	2 9 3
	6 3 0

Circle the number Three

0 9 2 1

4

3 5 3

8

4 2

7 5

10

2 6 3

Domino Addition

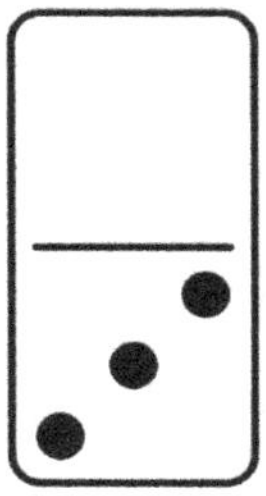 + 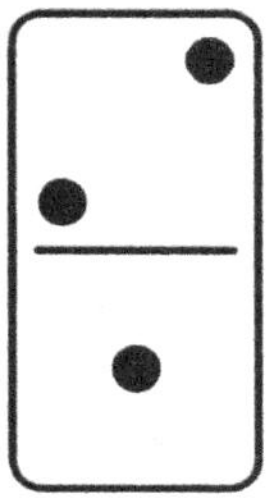 = ____

3 ____ 3 ____

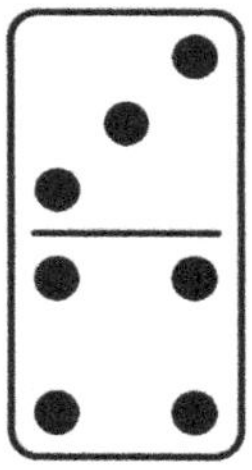 +  = ____

____ ____

 + 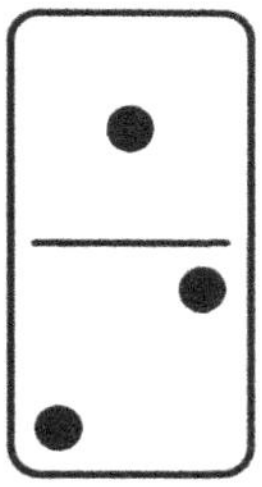 = ____

____ ____

 + 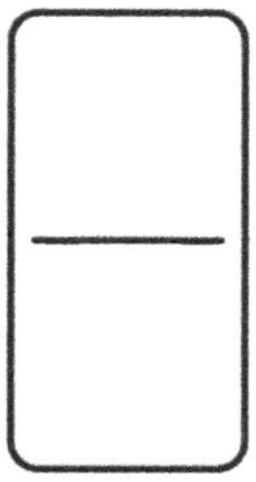 = ____

____ ____

Subtraction

3 − 2 = ____

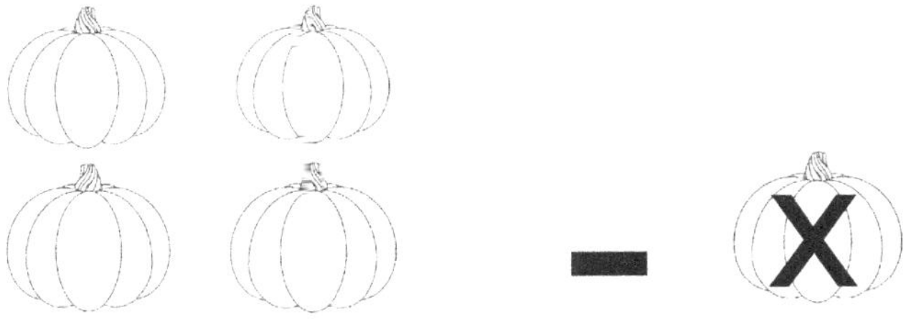

____ − ____ = ____

____ − ____ = ____

____ − ____ = ____

How many circles have waves
Circle and say the correct number

0 1 2 3 4 5 6 7 8 9 10

Four

Color the number Four on the right

Color the number Four on the Left

Color the number Four in the middle

Circle the correct number

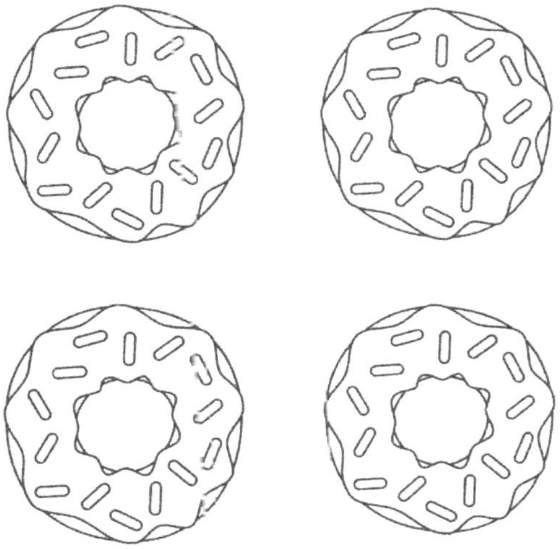

0 1 2 3 4

4 3 2 1 0

0 1 2 3 4

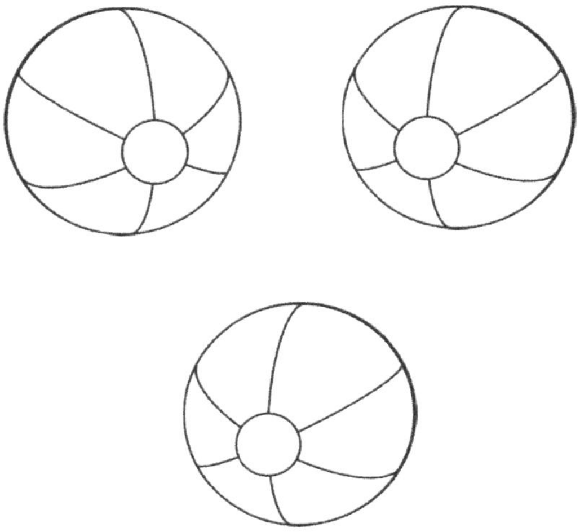

4 3 2 1 0

Domino Addition

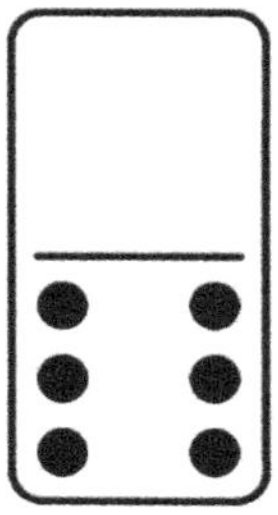 + 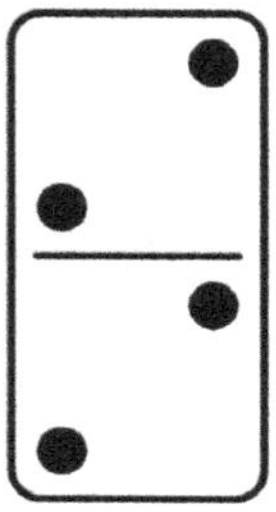 = ______

6 ______ 4 ______

 + 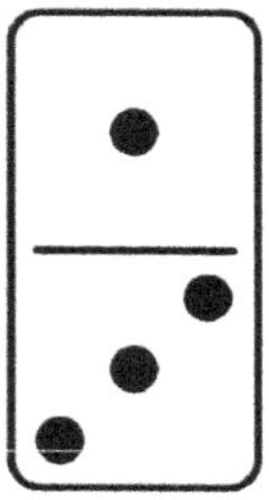 = ______

______ ______

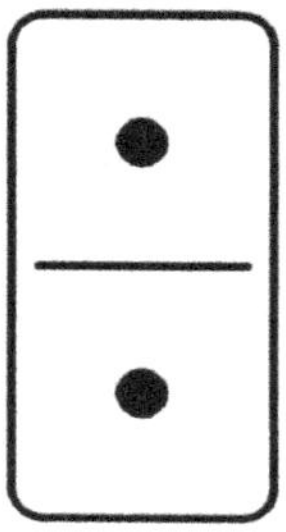 + 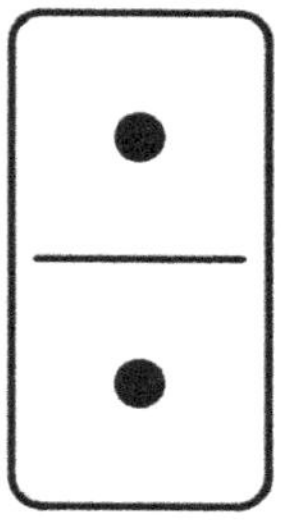 = ______

______ ______

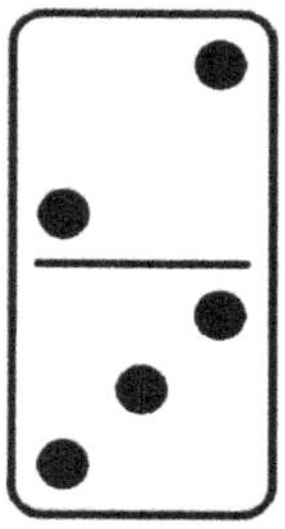 + 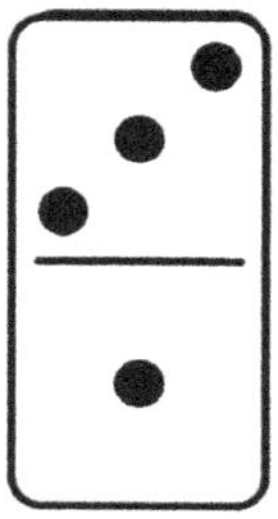 = ______

______ ______

Subtraction

4 1

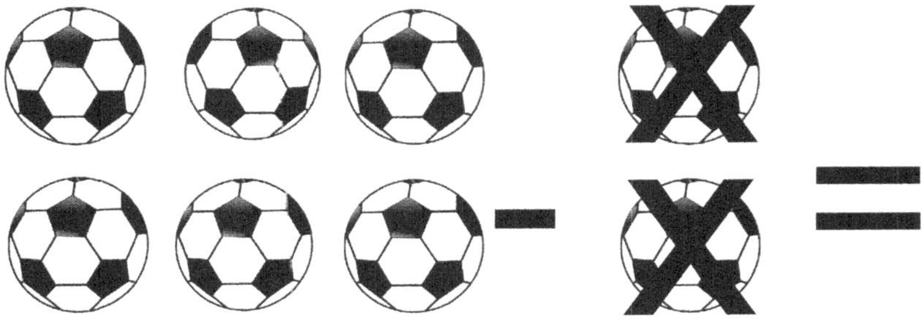

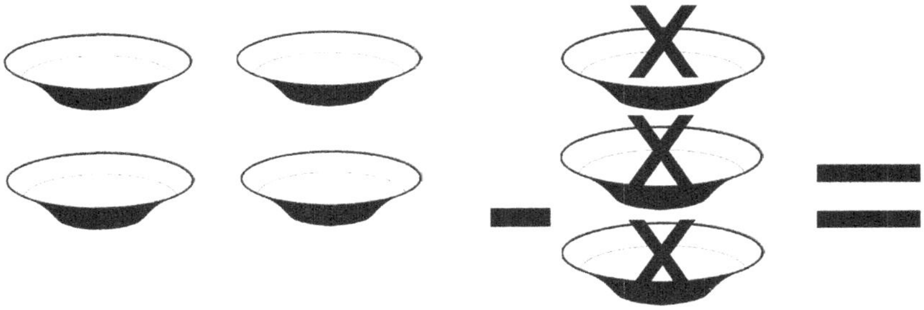

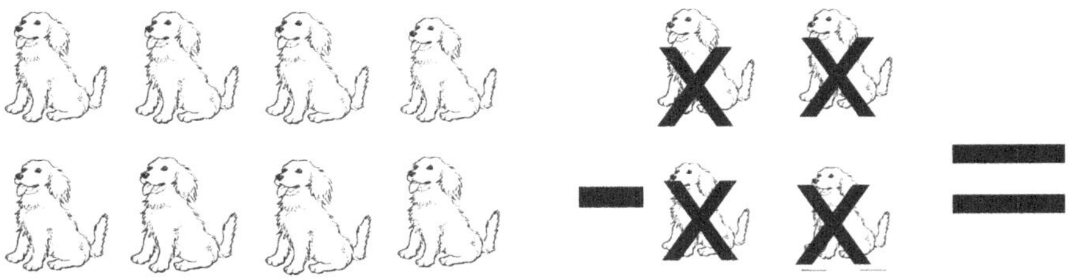

Color the triangles blue with the number 4

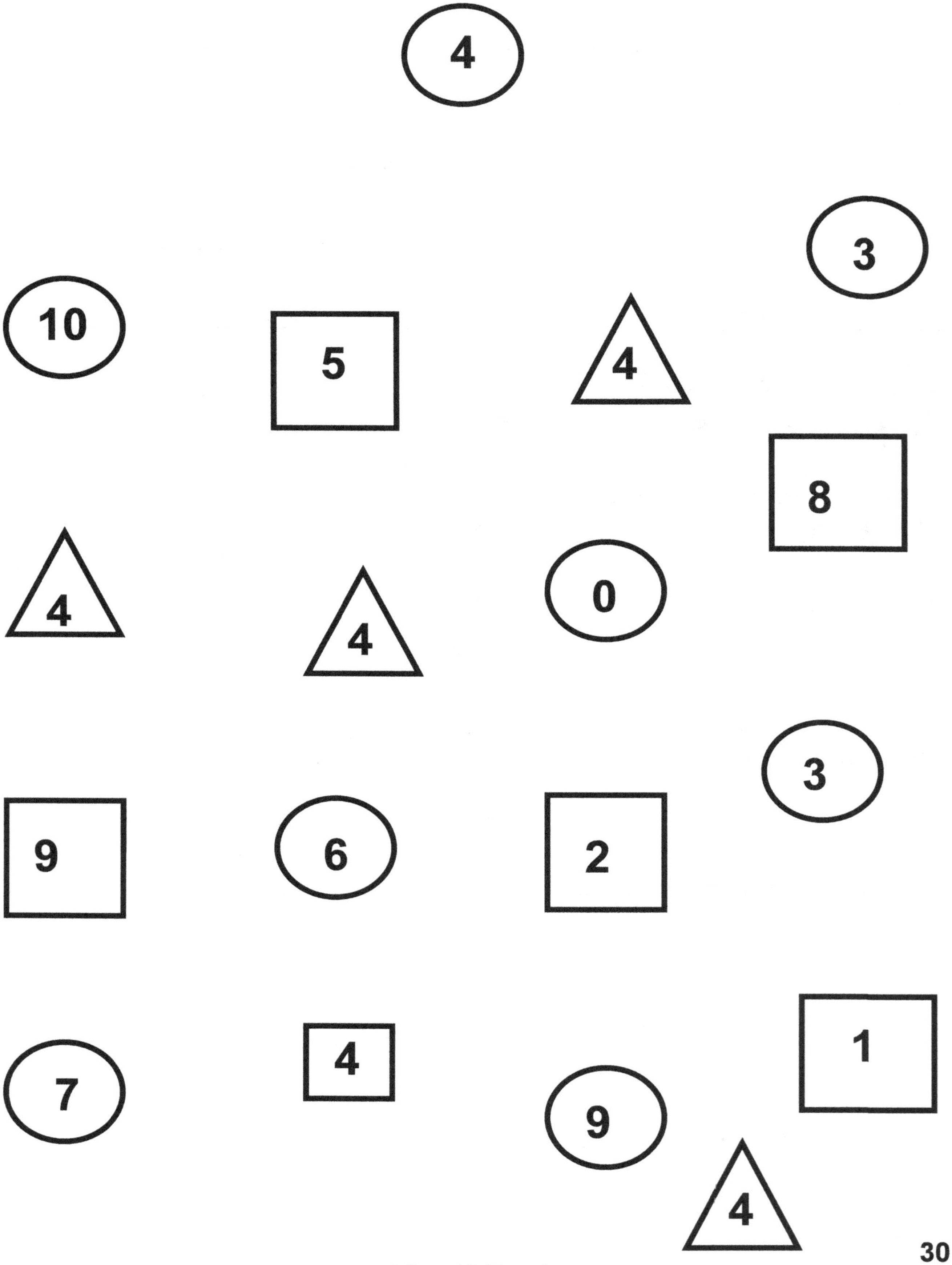

Five

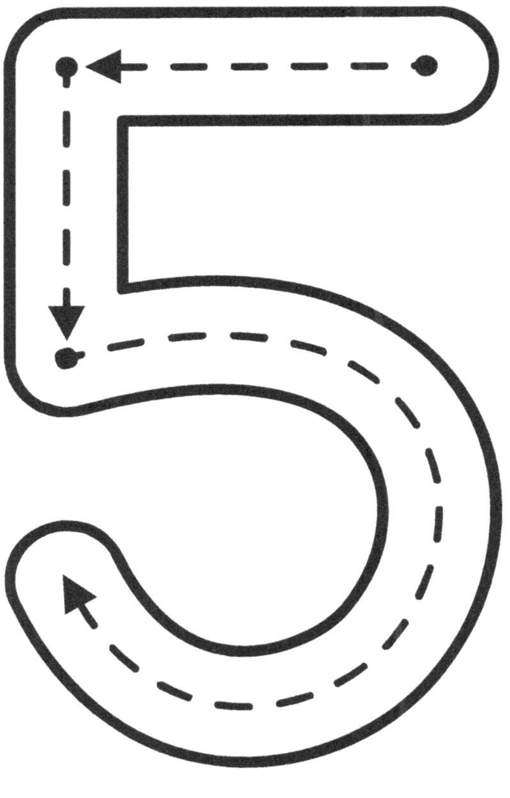

Color the largest 5 in each row

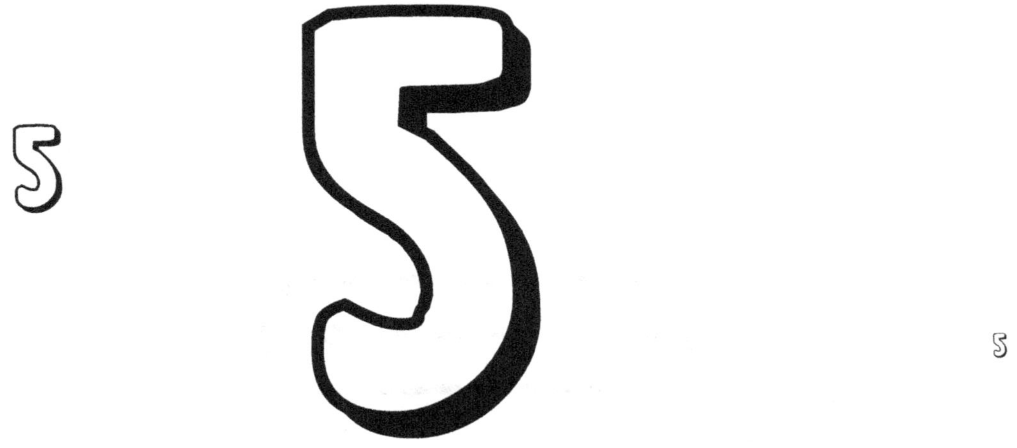

Circle the number Five

Domino Addition

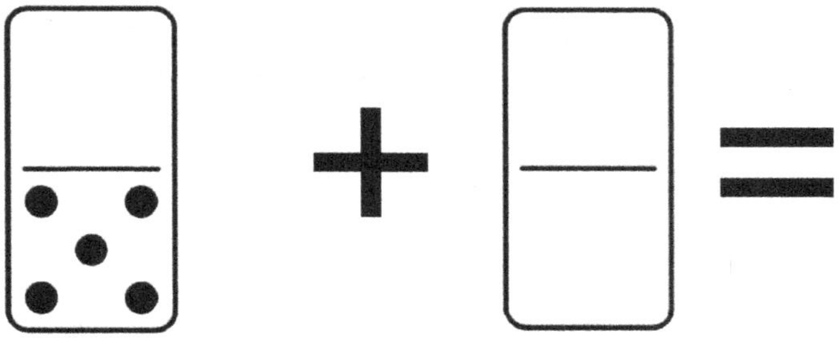

5 + 0 = ___

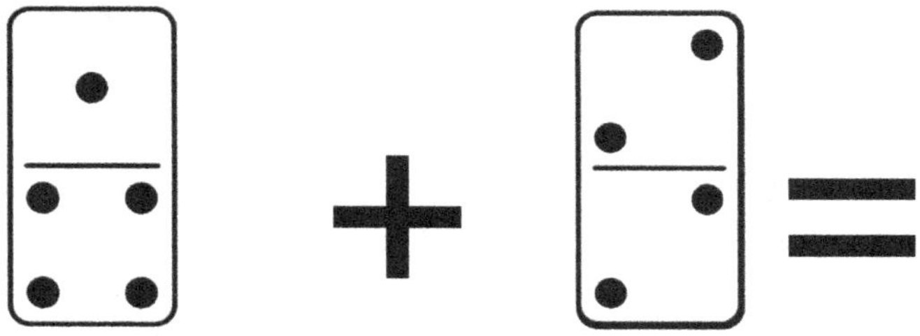

___ + ___ = ___

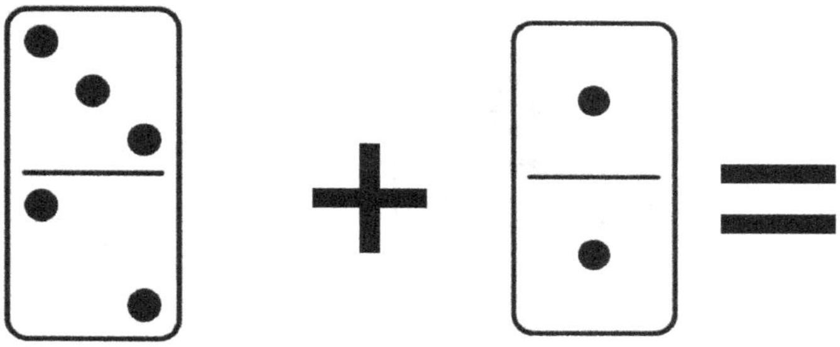

___ + ___ = ___

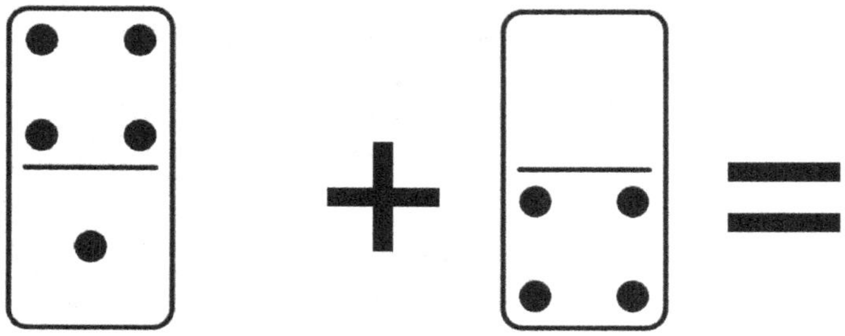

___ + ___ = ___

Subtraction

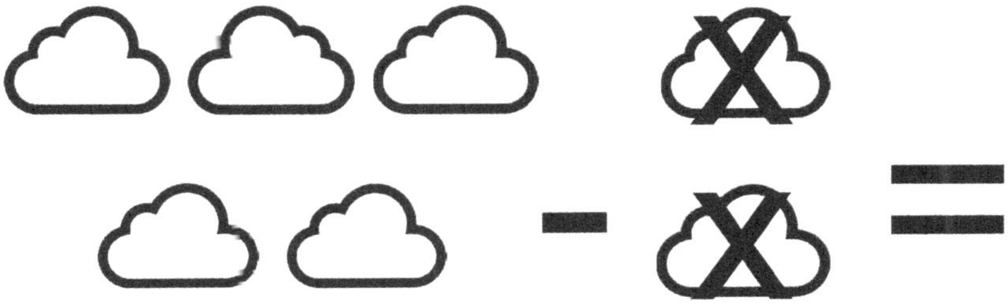

5   2

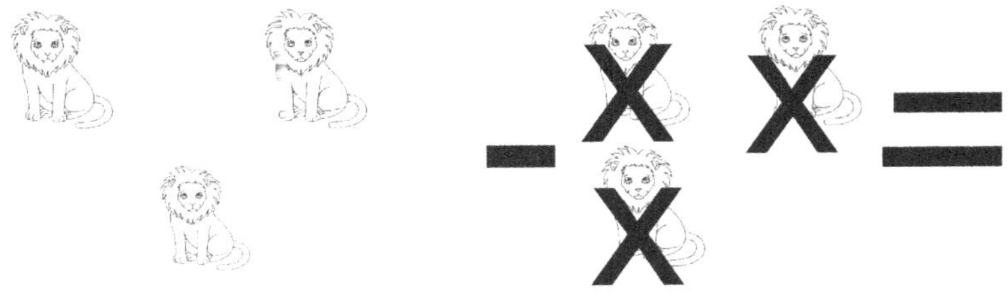

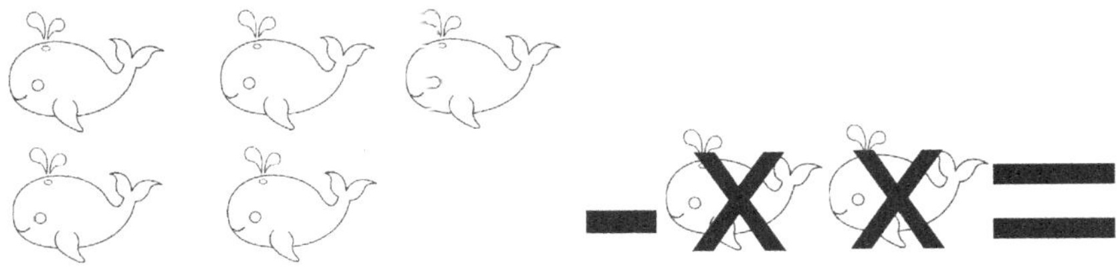

Circle what comes next

0 1 2 3 4 __	5 6
10 9 8 7 6 __	0 5
0 1 __	3 2
2 4 6 8 __	10 5

Six

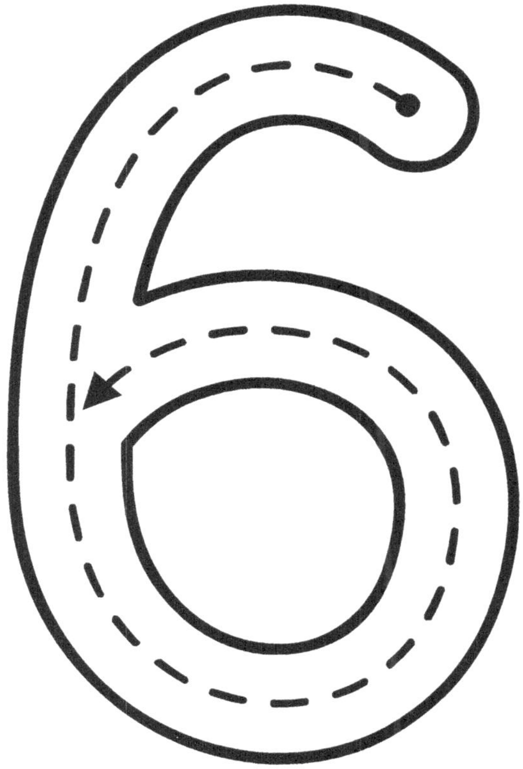

Put a square around the number 6

6

0 6 4

8

10

5 3

7

1

2

9 6

7

6 1

4

Circle the correct number of teddy bears

2	
0	
6	
3	
5	

Domino Addition

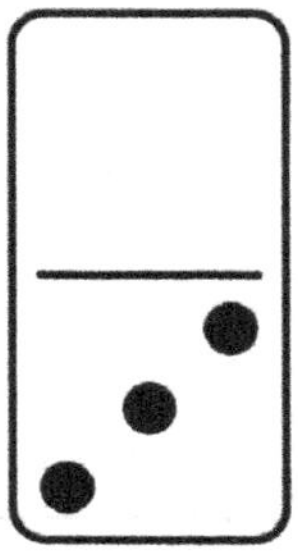 +  = ____

3 ____ 3 ____

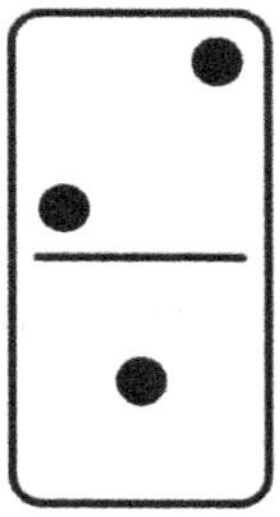 + 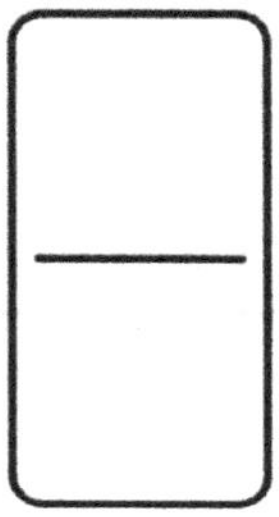 = ____

____ ____

 +  = ____

____ ____

 + 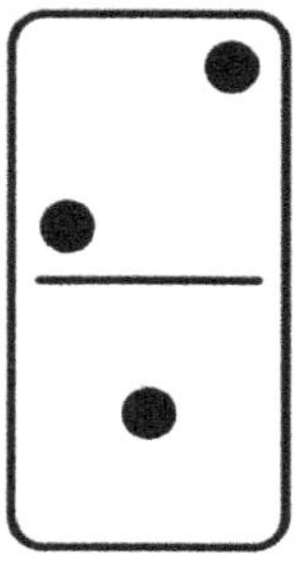 = ____

____ ____

Subtraction

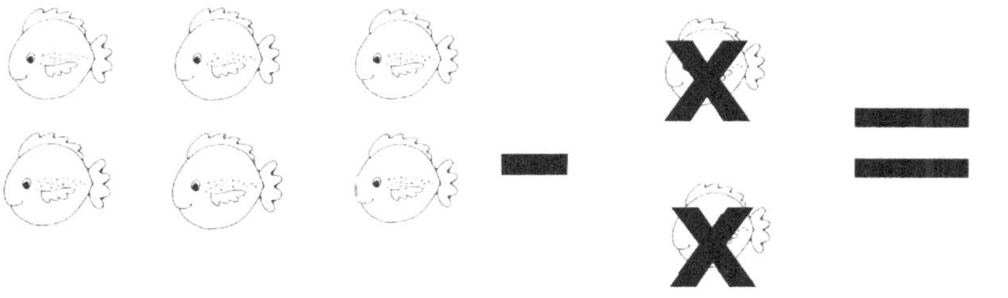

6 2

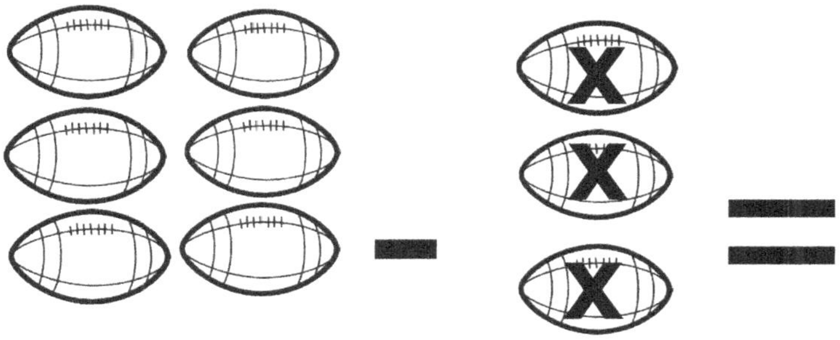

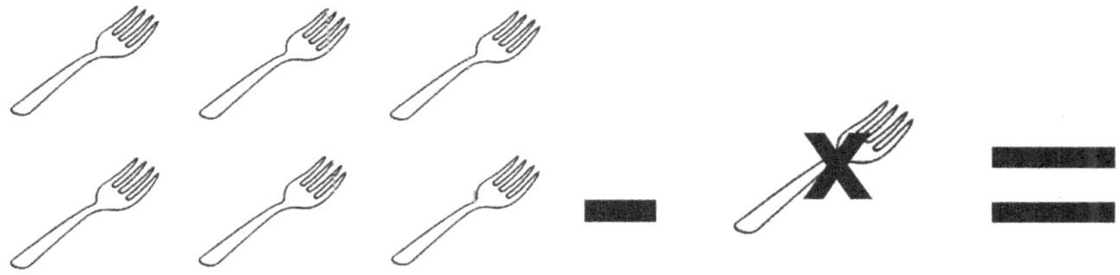

Color the ovals with the number Six

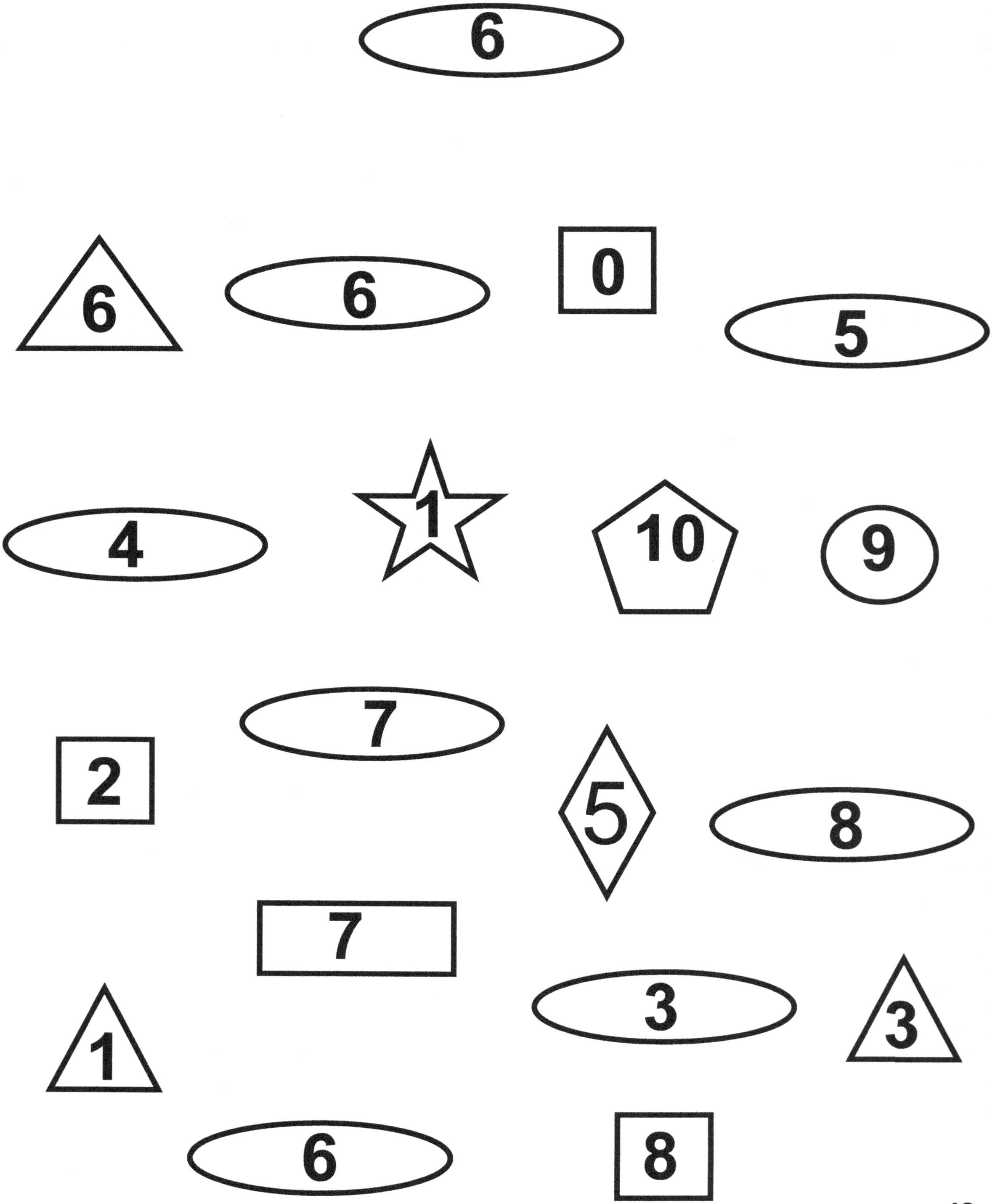

Seven

Color the bells with the number 7

Draw seven apples on the tree.

Domino Addition

 + 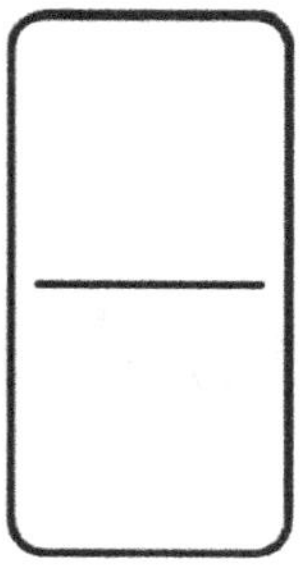 = ____

7 ____ 0 ____

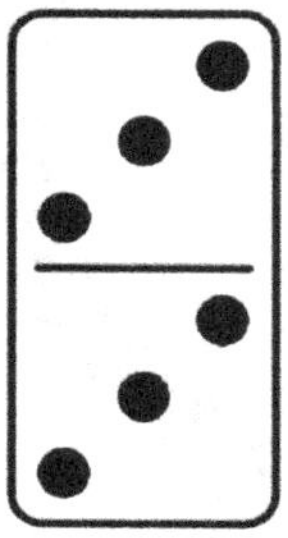 + 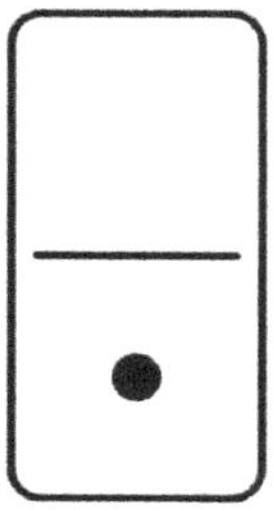 = ____

____ ____

 + 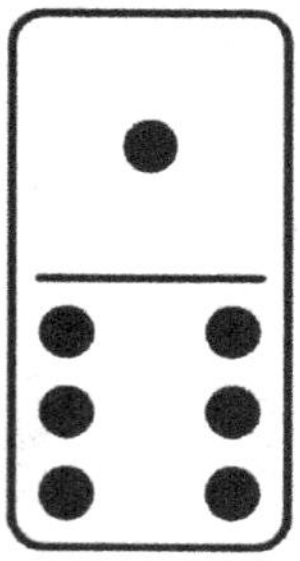 = ____

____ ____

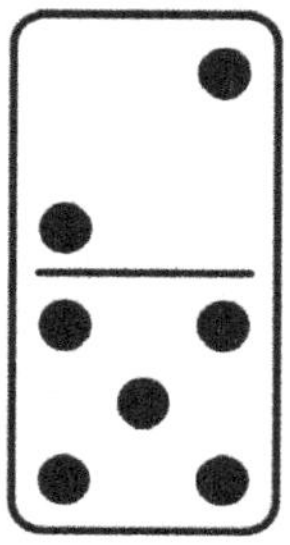 + 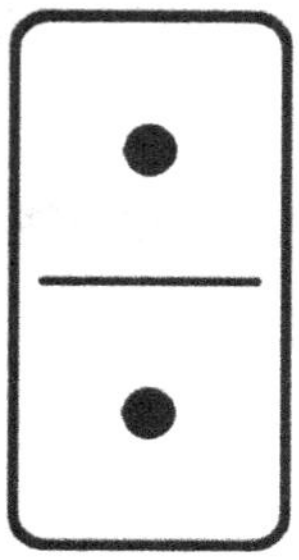 = ____

____ ____

Subtraction

 −  = ______

7 ______ 1 ______

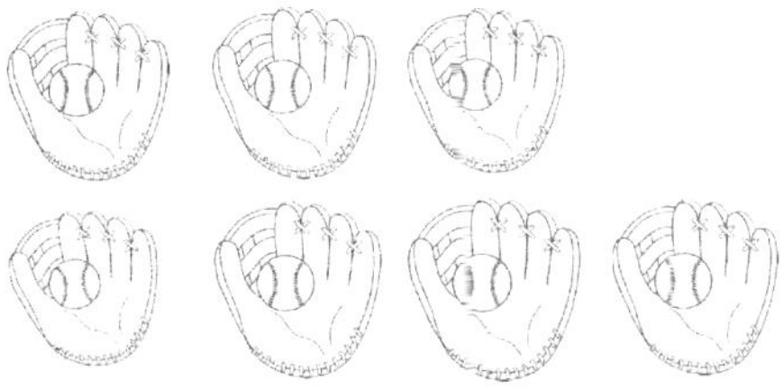 −  = ______

______ ______

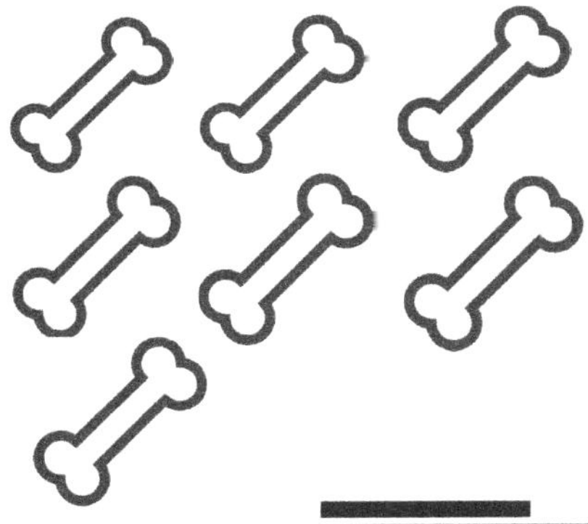 −  = ______

______ ______

 −  = ______

______ ______

Draw a line to match the numbers

1	4
6	6
4	1
10	0
0	10

Eight

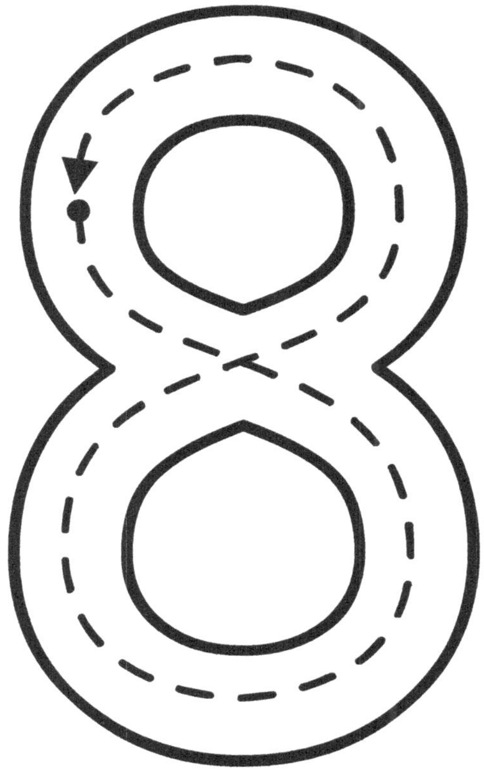

Circle the number 8

8 5

7

1

8

2

8 9

0

10

5

9

4

8

3

6

8

Draw eight flowers in the pot

Domino Addition

 +  = ______

3 ______ 5 ______

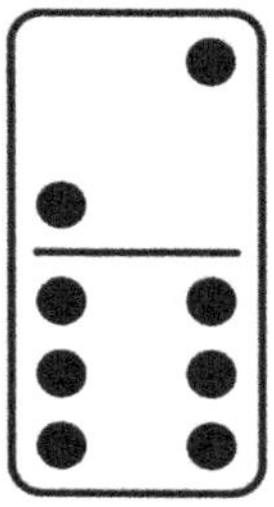 + 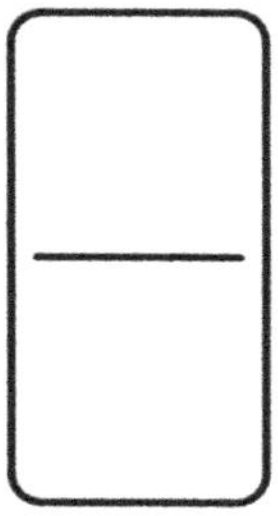 = ______

______ ______

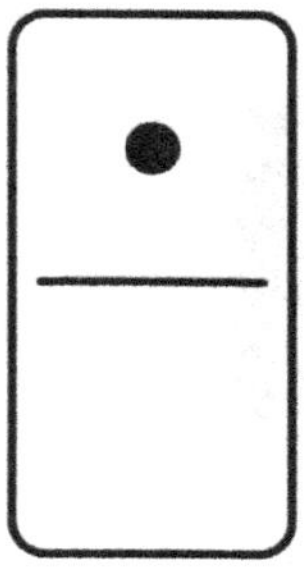 + 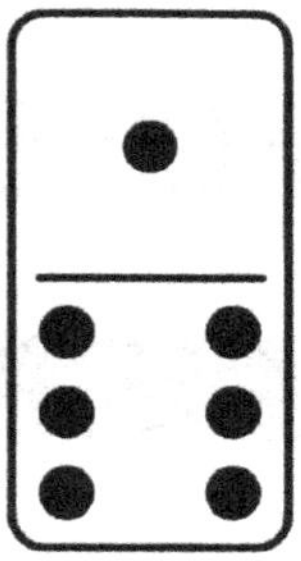 = ______

______ ______

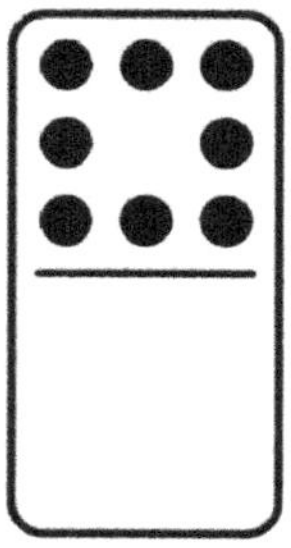 + 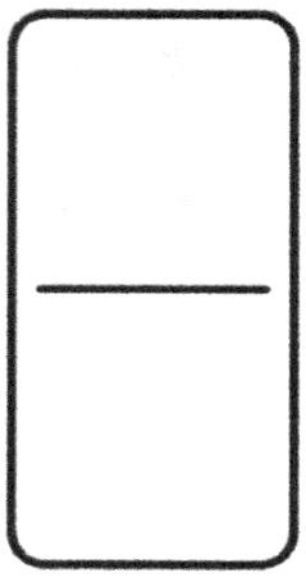 = ______

______ ______

Subtraction

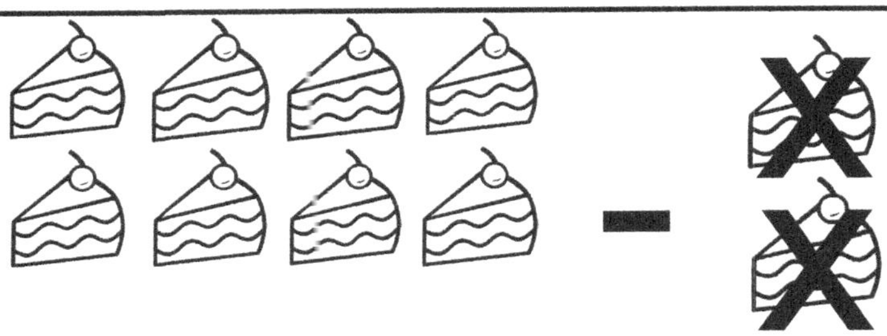

− =

8 ____ 2 ____ = ____

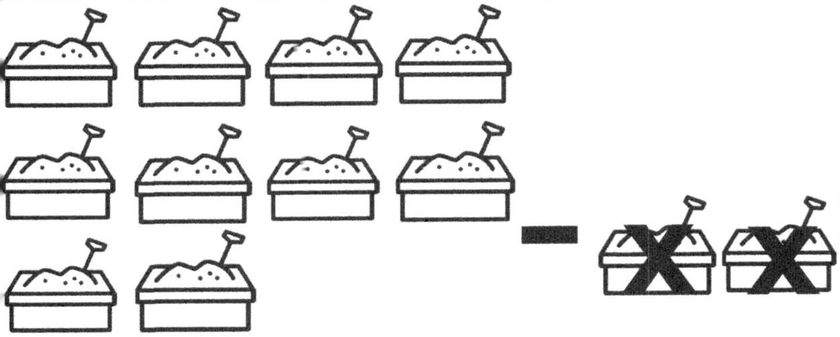

− =

____ ____ = ____

− =

____ ____ = ____

− =

____ ____ = ____

Write the missing numbers

0		2	
			4
			6

Nine

Color each pentagon with the number nine

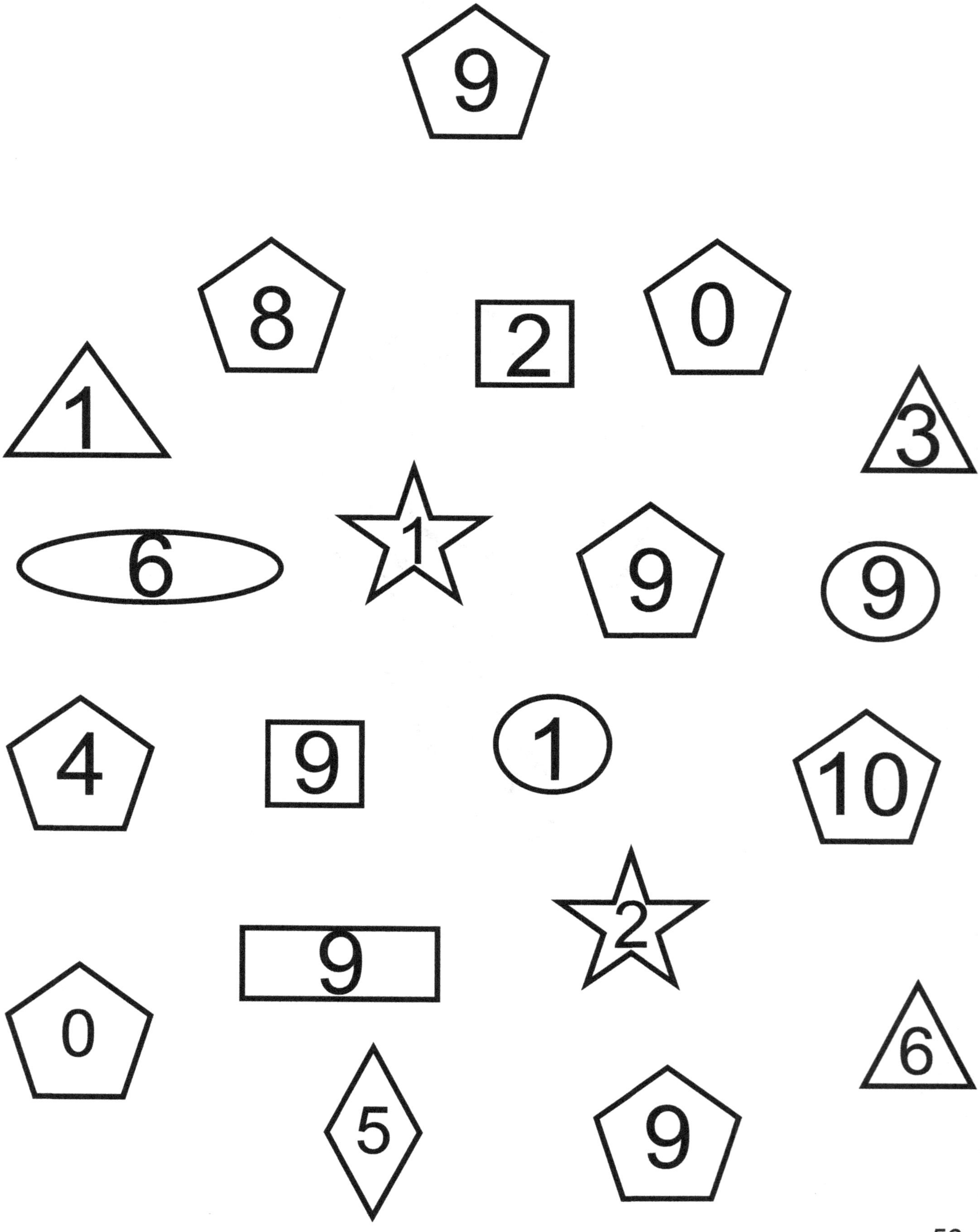

Circle 9 half moons

Domino Addition

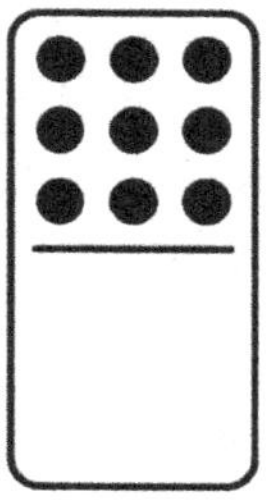 + 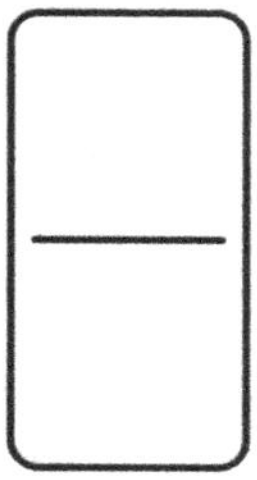 = ______

9 ______ 0 ______

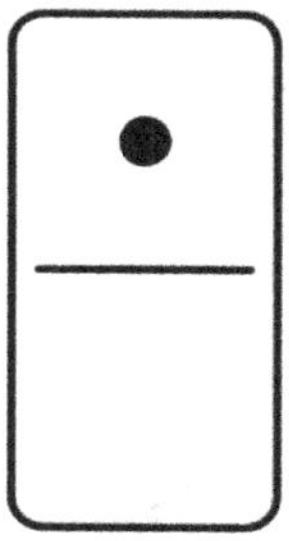 + 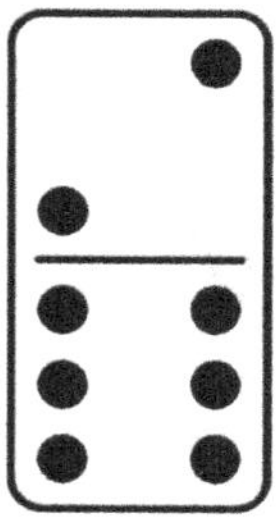 = ______

______ ______

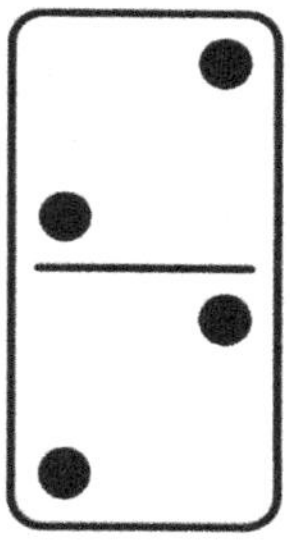 + 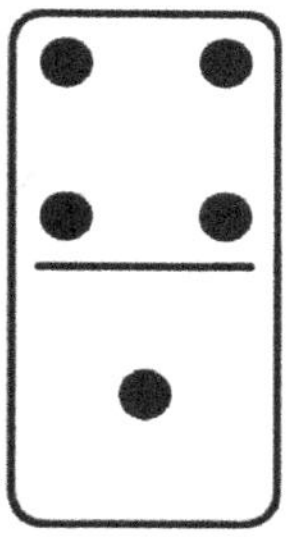 = ______

______ ______

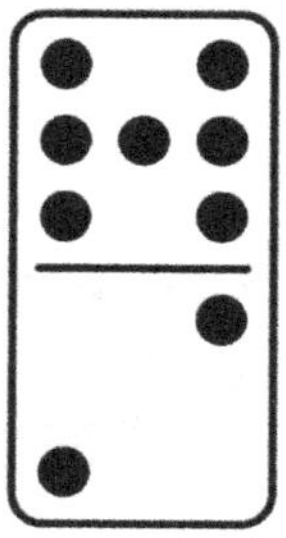 + 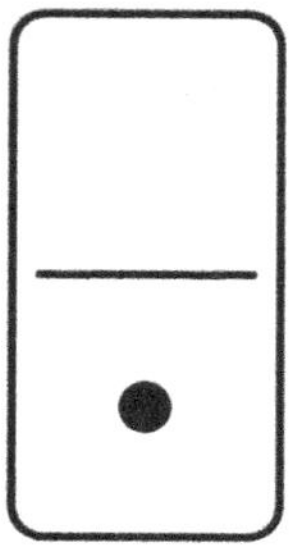 = ______

______ ______

Subtraction

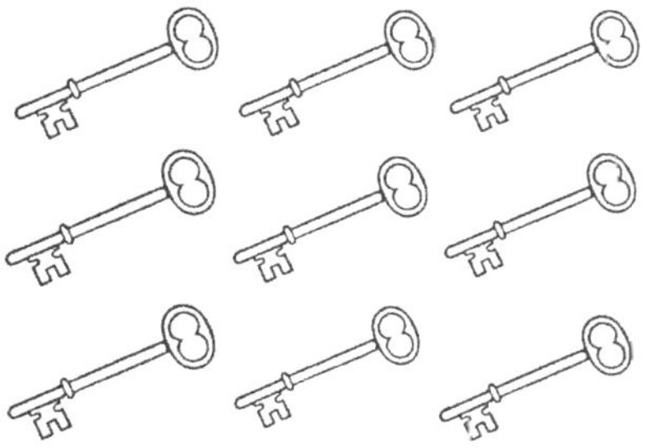 -  = ________

9 ________ 1 ________

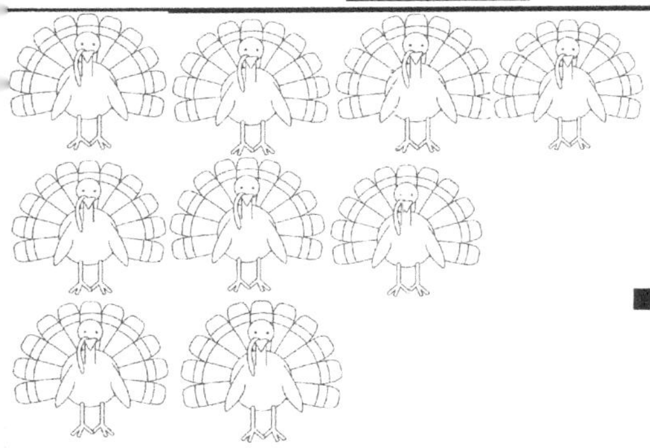 - 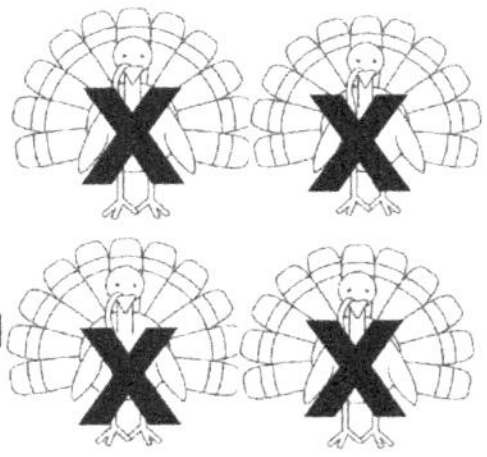 = ________

________ ________

 -  = ________

________ ________

 -  = ________

________ ________

Color 9 blocks

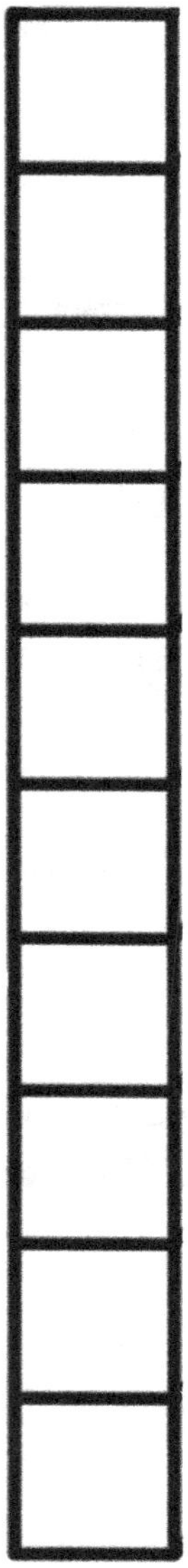

Ten

Write the number 10 on top of the table

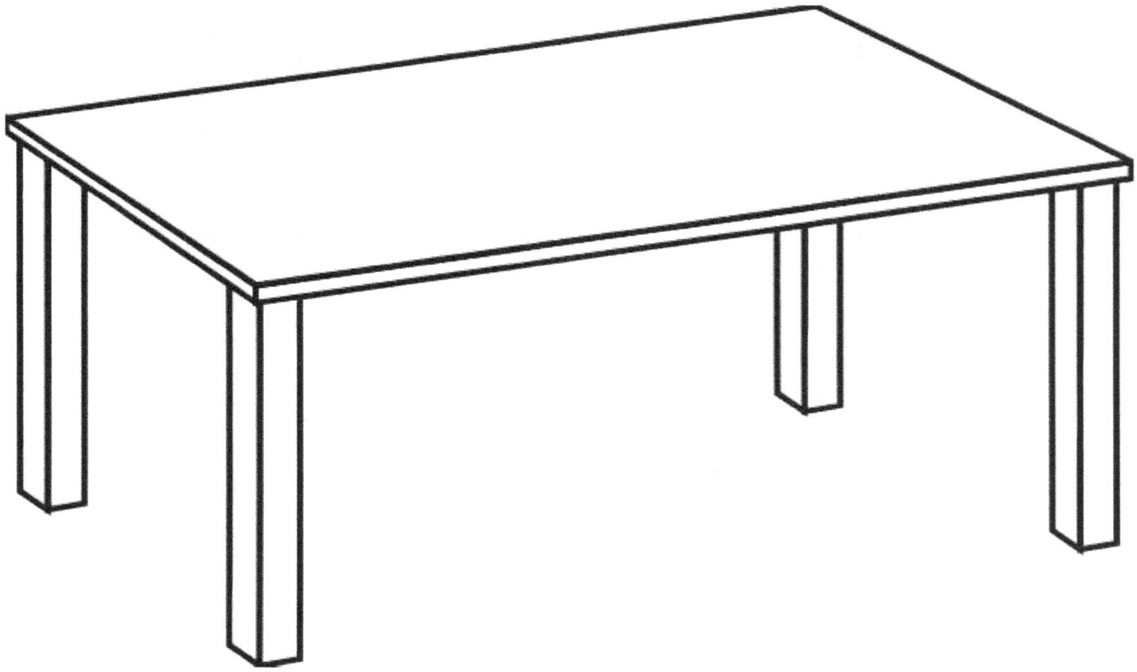

Color the number **ten**

Domino Addition

 +  = ____

5   5

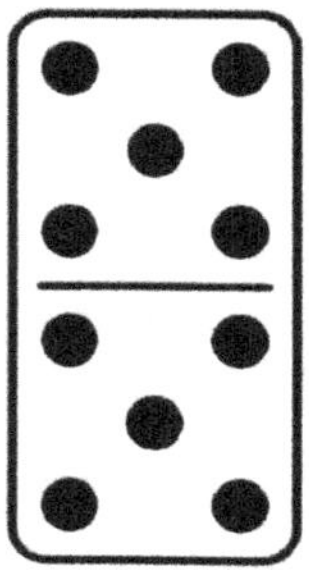 + 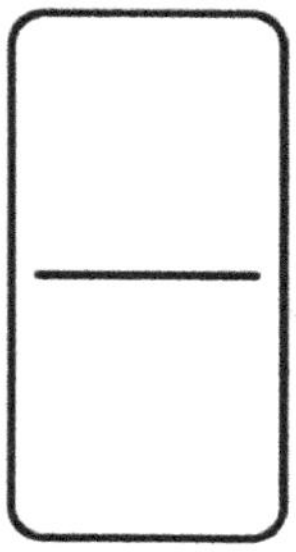 = ____

____   ____

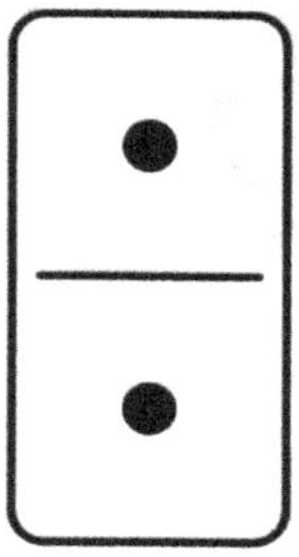 + 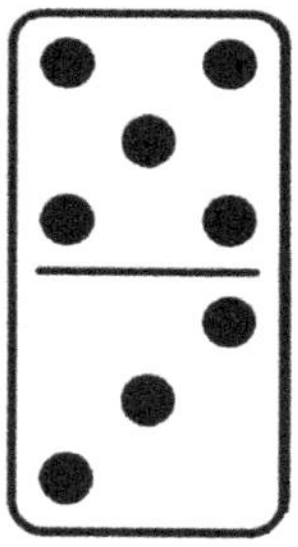 = ____

____   ____

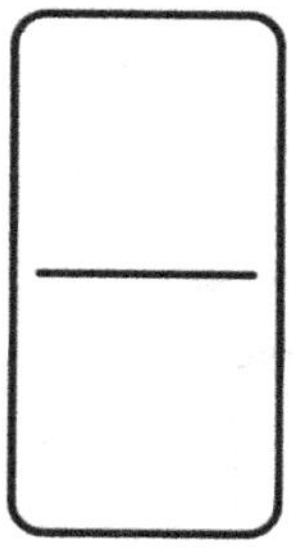 + 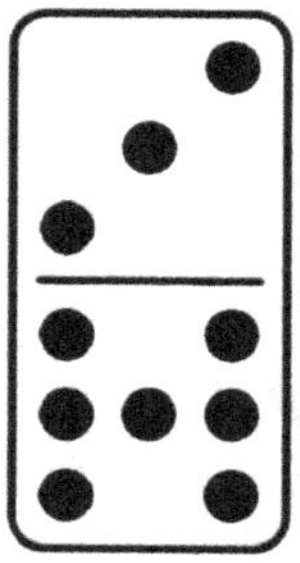 = ____

____   ____

Subtraction

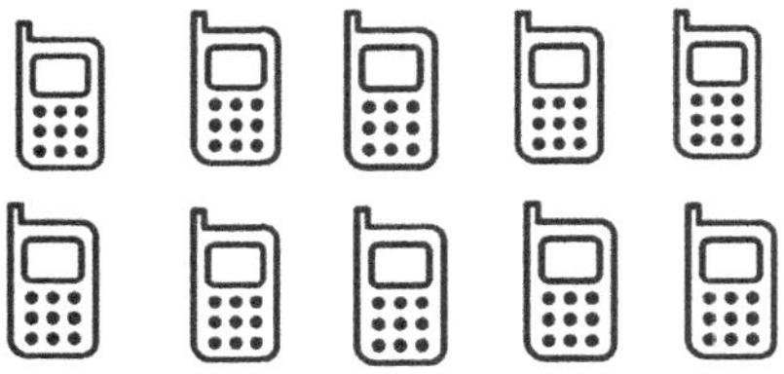 −  = ______

10 ______ 2 ______

 −  = ______

______ ______

 −  = ______

______ ______

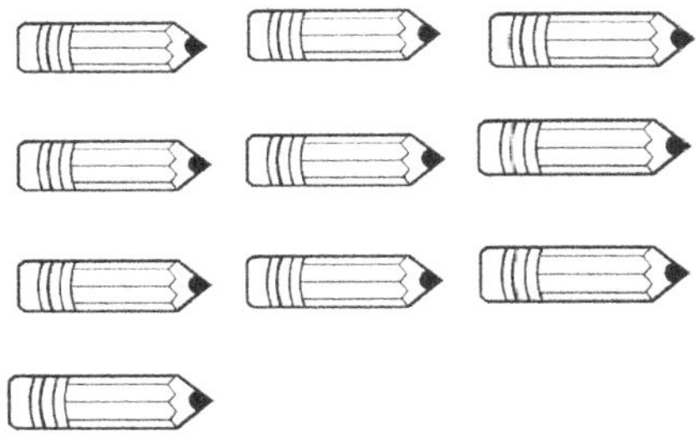 − 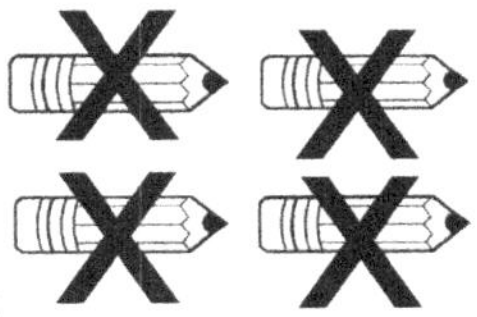 = ______

______ ______

Connect the dots

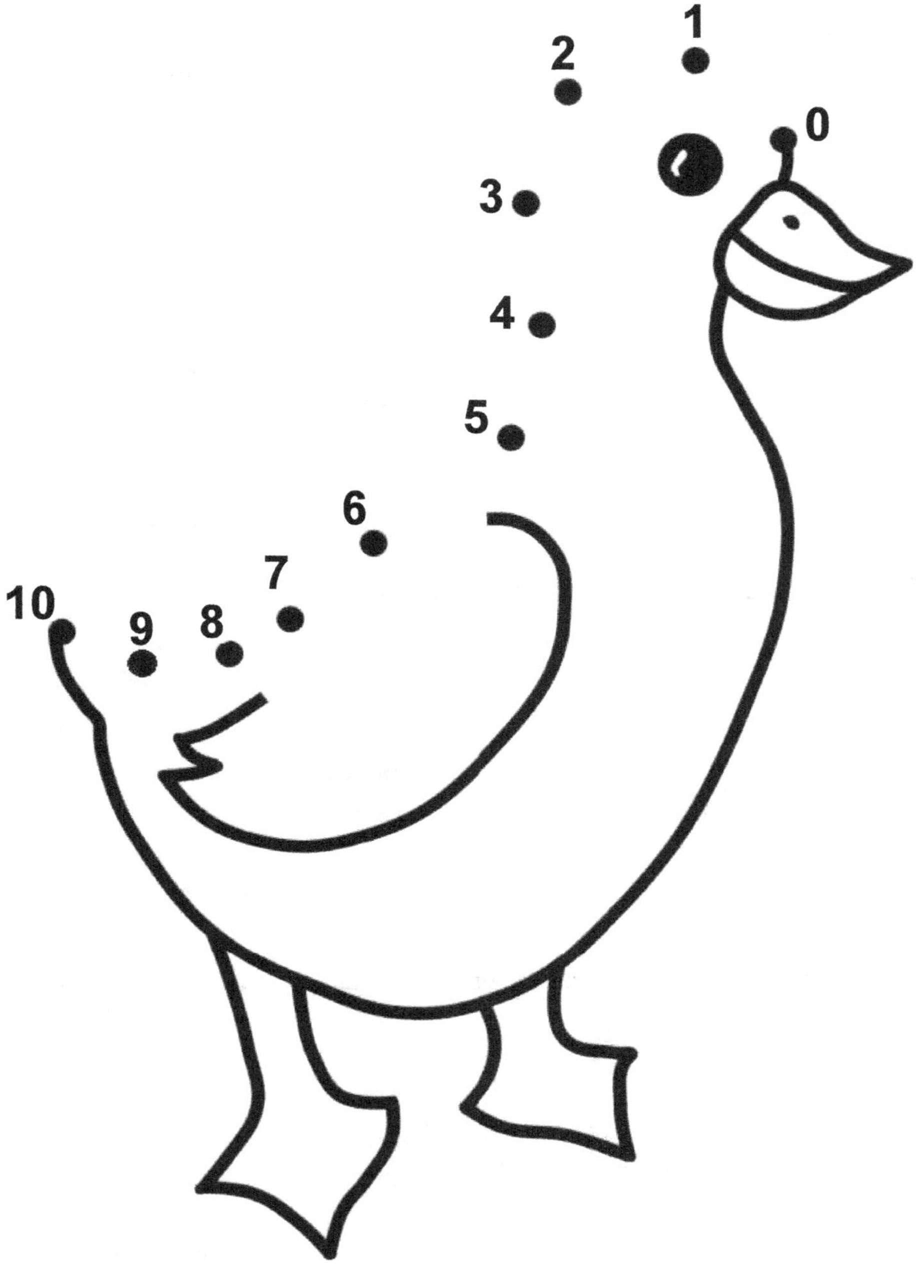

Knowledge Review Section

Trace the numbers 0-10

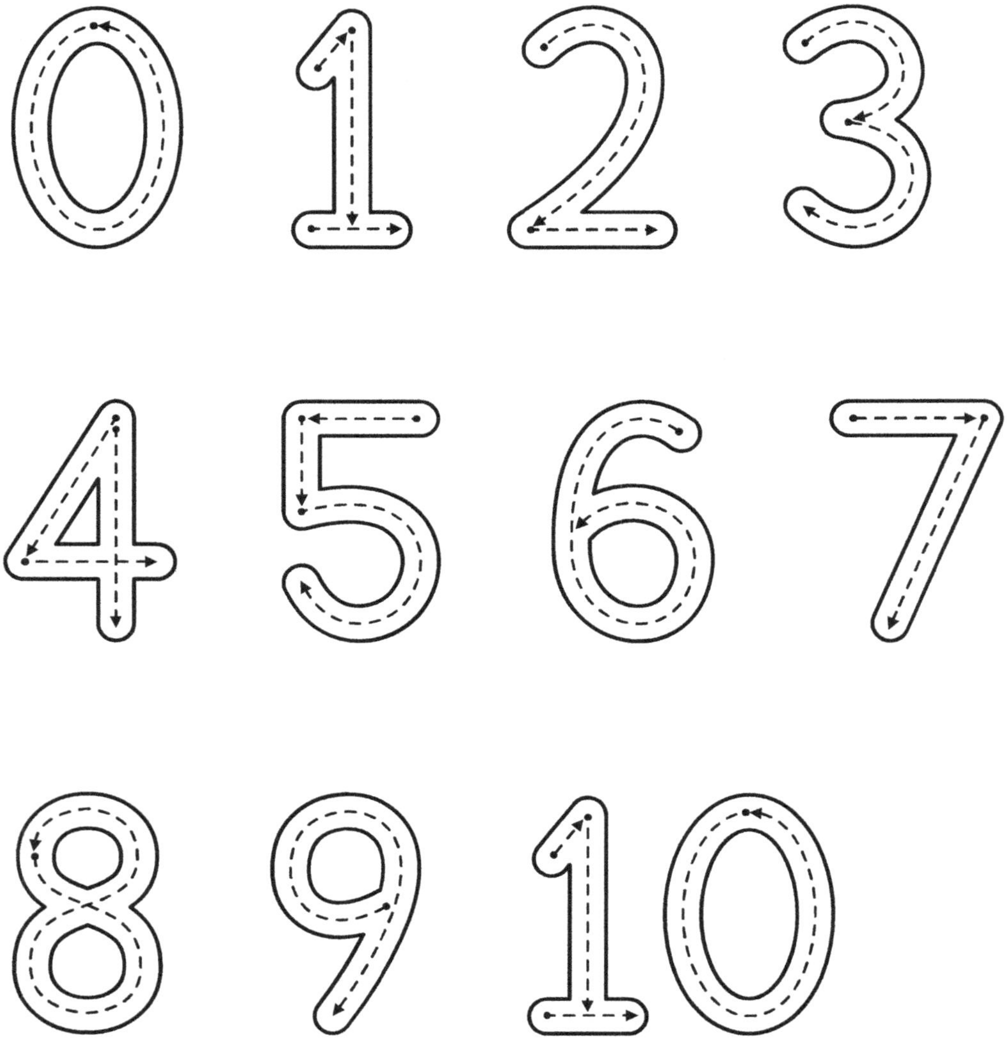

Addition Worksheets

Name____________________

$0 + 0 = \square$	$1 + 1 = \square$	$2 + 2 = \square$
$3 + 3 = \square$	$4 + 4 = \square$	$5 + 5 = \square$
$1 + 2 = \square$	$0 + 3 = \square$	$7 + 2 = \square$

Name____________________

1 + 0 = ☐	7 + 3 = ☐	9 + 1 = ☐
4 + 3 = ☐	8 + 2 = ☐	1 + 5 = ☐
1 + 7 = ☐	5 + 0 = ☐	9 + 0 = ☐

Name___________________

3 + 4 = ☐	1 + 8 = ☐	5 + 2 = ☐
6 + 0 = ☐	2 + 7 = ☐	0 + 1 = ☐
4 + 1 = ☐	6 + 2 = ☐	10 + 0 = ☐

Name____________________

6 + 3 = ☐	2 + 8 = ☐	3 + 1 = ☐
8 + 0 = ☐	8 + 2 = ☐	5 + 3 = ☐
7 + 1 = ☐	2 + 4 = ☐	0 + 6 = ☐

Substraction Worksheet

Name____________________

1 - 0 ☐	2 - 1 ☐	8 - 4 ☐
9 - 1 ☐	8 - 2 ☐	4 - 4 ☐
6 - 1 ☐	7 - 0 ☐	5 - 2 ☐

Name____________________

1 - 1 ☐	0 - 0 ☐	2 - 2 ☐
6 - 2 ☐	10 - 6 ☐	7 - 4 ☐
4 - 2 ☐	5 - 1 ☐	7 - 7 ☐

Name____________________

9 − 5 = ☐	5 − 3 = ☐	3 − 3 = ☐
6 − 0 = ☐	3 − 1 = ☐	7 − 1 = ☐
9 − 4 = ☐	10 − 1 = ☐	4 − 0 = ☐

Name________________

7 - 2 = ☐	2 - 0 = ☐	5 - 5 = ☐
10 - 8 = ☐	8 - 0 = ☐	3 - 2 = ☐
4 - 1 = ☐	8 - 1 = ☐	6 - 6 = ☐

Bonus Topics

Greater than or Less than Worksheets

Name___________________

< Less than **> Greater than**

0	◯	3
1	◯	10
7	◯	1
8	◯	3

Name____________________

< Less than **> Greater than**

10	◯	6
3	◯	8
5	◯	1
7	◯	4

Name___________________

< Less than **> Greater than**

5	○	7
9	○	0
8	○	6
3	○	1

Name_______________

< Less than **> Greater than**

1	◯	5
9	◯	7
10	◯	0
0	◯	7

Fractions Worksheets

Name________________________

Halves

Color $\frac{1}{2}$ of each shape.

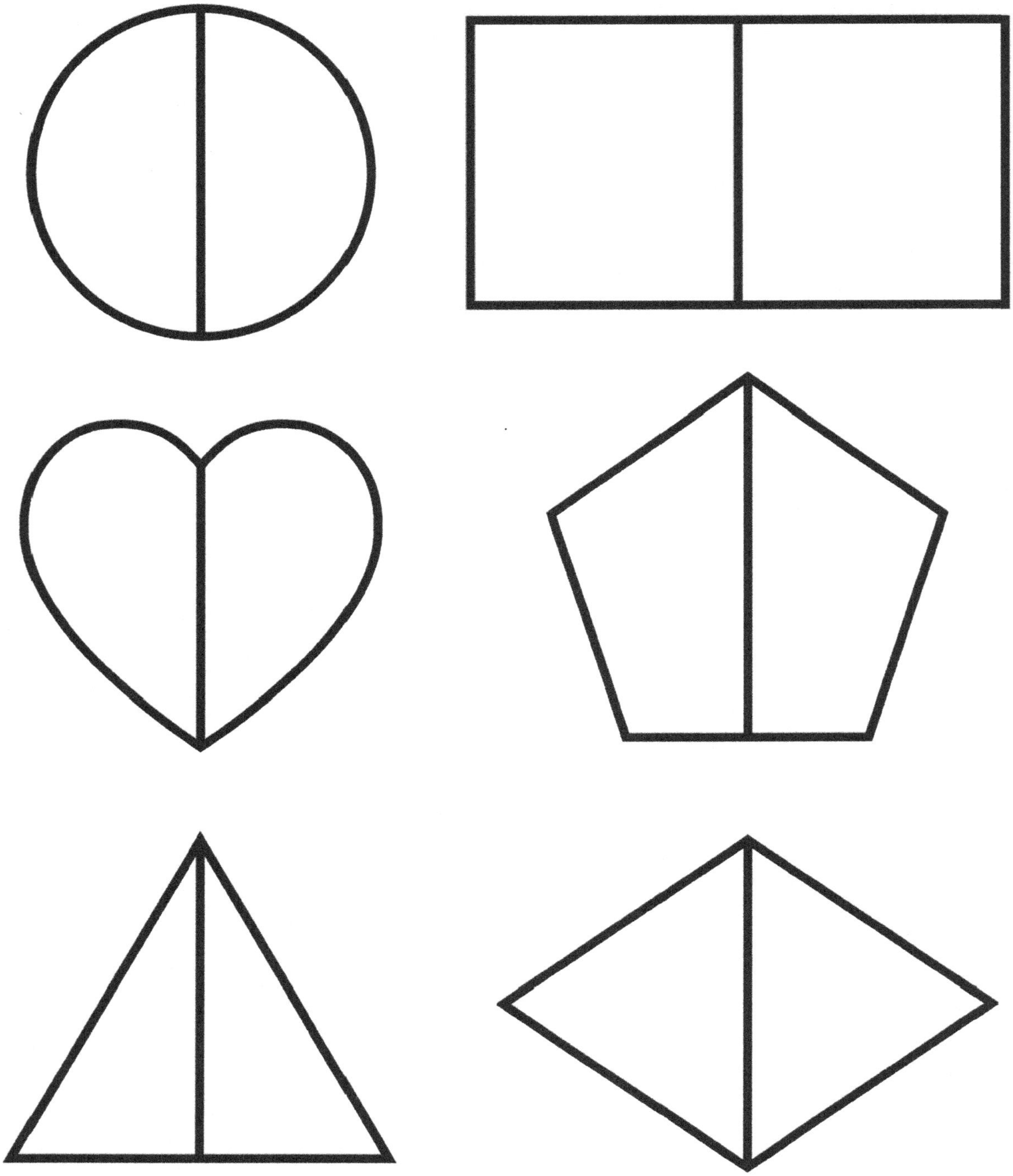

Name____________________

Color the fraction circle

$\frac{1}{2}$ =

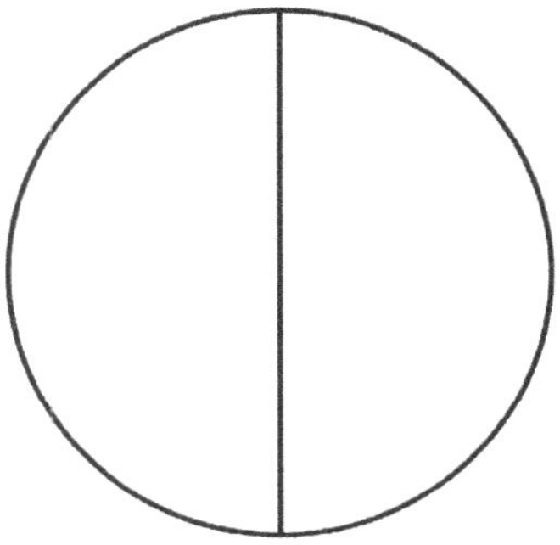

$\frac{1}{3}$ =

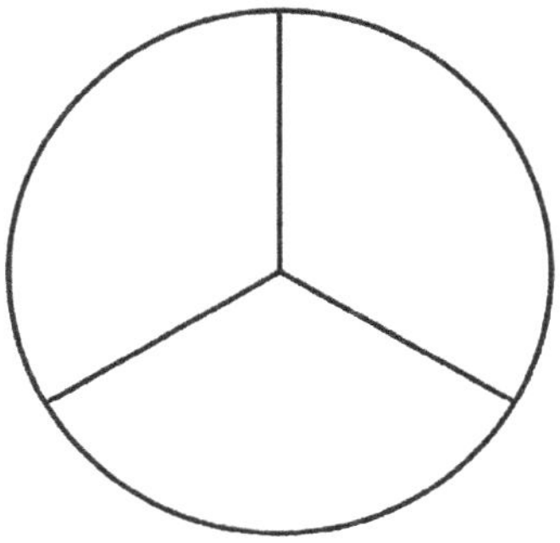

$\frac{1}{4}$ =

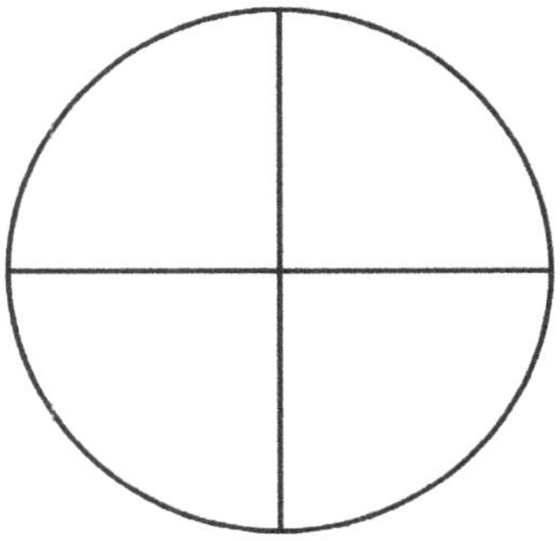

$\frac{1}{5}$ =

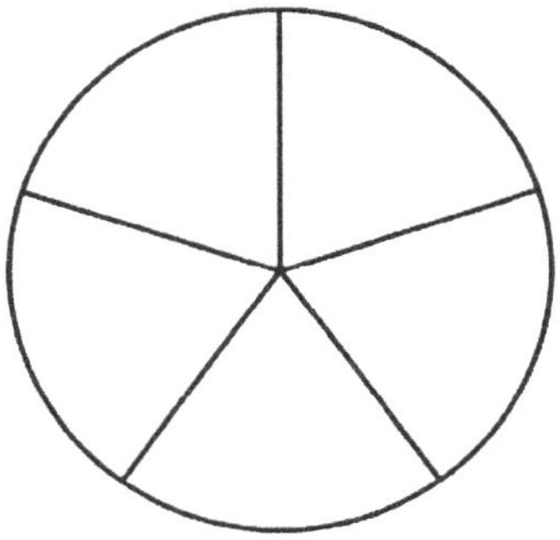

$\frac{1}{6}$ =

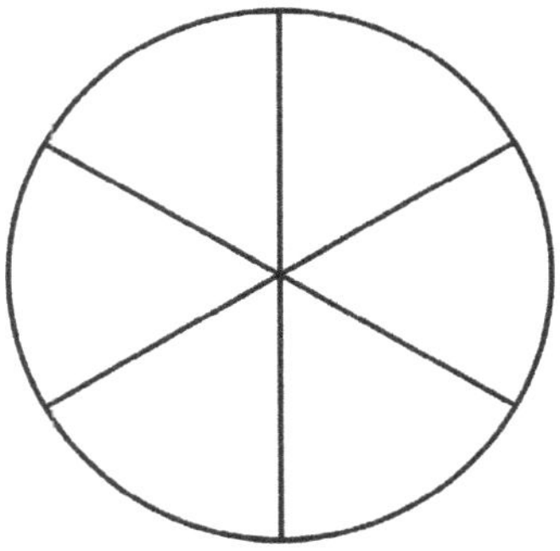

$\frac{1}{7}$ =

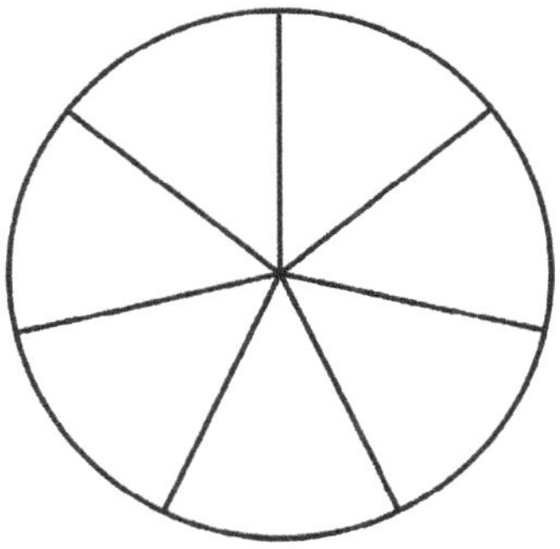

$\frac{1}{8}$ =

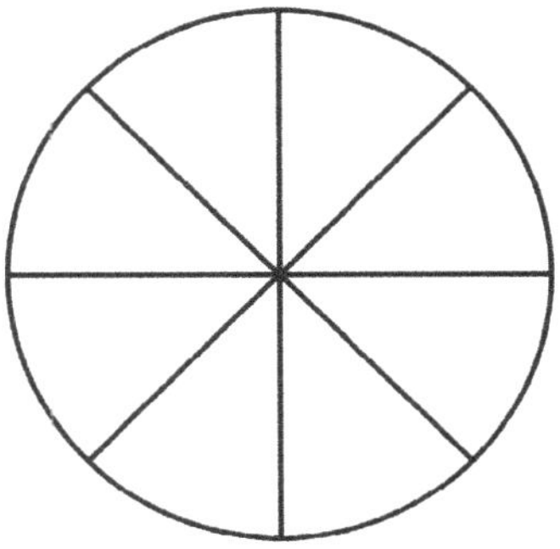

$\frac{1}{9}$ =

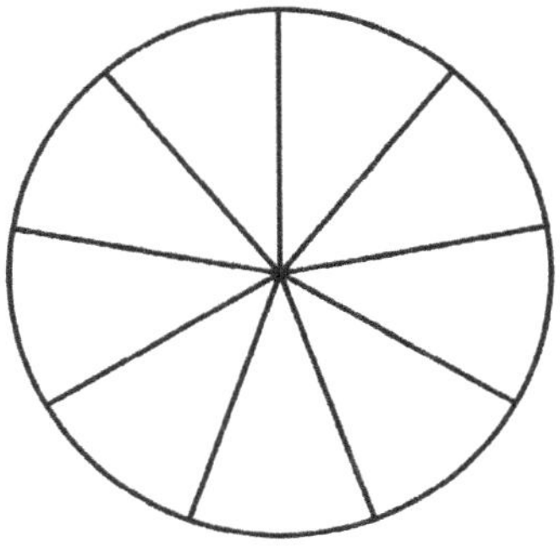

Telling time by the hour Worksheet

Name________________

Telling time by the hour

Match the digital clock and analog clock that shows the same time.

Digital Clocks | Analog Clocks

Printing Publishing Palace

Superstar Award

____

is here by congratulated on completing

Preschool Math Concepts for Beginner Learners

Presented on ____________________ Parent or Guardian: ____________________

Made in the USA
Monee, IL
08 November 2026

67971102R00052